PRAISE FOR *EL NIÑO*

"Well-paced and realized...a solid work of scientific reportage...a fine contribution to popular-science literature.... Nash is very good at explaining highly complex science and at engaging her readers."
— *Kirkus Reviews*

"For those who want to understand the complexities and the grave dangers of a warming world, EL NIÑO provides the best analog. And this fine volume provides the best account of those earth-shaking cycles. It's as important a part of emergency preparedness as extra water and new batteries."
—Bill McKibben, author of *The End of Nature* and *Long Distance*

"The story of the perfect storm system.... Fascinating."
— *Publishers Weekly*

"A memorable journey...fascinating...thought-provoking.... Nash melds compelling stories of scientific detective work with a gripping account of what we know about the workings of 'the Christmas child.'"
—Brian Fagan, professor, University of California,
Santa Barbara, and author of
The Little Ice Age: How Climate Made History 1300–1850

"Vivid.... Nash puts a human face on the phenomenon."
— *Library Journal*

"Educational and well-researched...provides coherent, digestible explanations of enormous and complex global events."
— *San Diego Union-Tribune*

EL NIÑO

UNLOCKING THE SECRETS OF THE MASTER WEATHER-MAKER

J. MADELEINE NASH

WARNER BOOKS

An AOL Time Warner Company

Copyright © 2002 by J. Madeleine Nash
All rights reserved.

Warner Books, Inc., 1271 Avenue of the Americas, New York, NY 10020

Visit our Web site at www.twbookmark.com.

 An AOL Time Warner Company

Printed in the United States of America
Originally published in hardcover by Warner Books, Inc.

First Trade Printing: March 2003

10 9 8 7 6 5 4 3 2 1

The Library of Congress has catalogued the hardcover edition as follows:

Nash, J. Madeleine.
    El Niño : unlocking the secrets of the master weather-maker / J. Madeleine Nash.
        p. cm.
    Includes index.
    ISBN 0-446-52481-6
        1. El Niño Current—Environmental aspects.   I. Title.

GC296.8.E4 N37 2001
551.6—dc21

2001045602

ISBN 0-446-67992-5 (pbk.)

Book design by Giorgetta Bell McRee

ATTENTION: SCHOOLS AND CORPORATIONS
WARNER books are available at quantity discounts with bulk purchase for educational, business, or sales promotional use. For information, please write to: SPECIAL SALES DEPARTMENT, WARNER BOOKS, 1271 AVENUE OF THE AMERICAS, NEW YORK, NY 10020

*For my mother, who taught me
to love the wind and the rain*

CONTENTS

GREENLAND

NORTH
AMERICA

*North
Pacific
Ocean*

*North
Atlantic
Ocean*

RIO NIDO,
CALIF

4 CORNERS

KIRITIMATI ISLAND
(Christmas Island)

PAITA, PERU

QUELCCAYA ICE CAP

SOUTH
AMERICA

*South
Pacific
Ocean*

*South
Atlantic
Ocean*

CHAPTER **ONE**

*But not alone has this Leviathan left his pre-Adamite traces
in the stereotype plates of nature, and in limestone and marl
bequeathed his ancient bust; but upon Egyptian tablets,
whose antiquity seems to claim for them an almost fossilif-
erous character, we find the unmistakable imprint of his fin.*
                                    HERMAN MELVILLE, *Moby Dick*

RIO NIDO

Gary LaCombe was exhausted. For a long moment he just
stood at the top of his driveway and stared into the
stream of mocha-colored water swirling around his feet.
"Maybe we can get the county to help us," said his wife, Phyllis,
who was just as tired as he was.

Around that time, from somewhere high above, came a series
of sharp reports that sounded a lot like rifle fire. Near the top of
the steep-sided canyon that angled up behind the house, Gary figured, big trees must be falling, probably Douglas firs whose shallow roots had lost their grip on waterlogged soil. Far and wide,
the scent that wafted from those broken evergreen boughs in-
vested the air with a mysterious fragrance. The odd thing was,
there was no wind. Trees usually fall, Gary reflected, when there's
wind.

It was late on Friday night, February 6, 1998. In northern California the winter-storm season was in full swing, and the La-Combes and their neighbors had been waging a nonstop struggle to keep culverts and creek beds clear of rocks, mud, and debris; otherwise these conduits would overflow and send water spilling into yards, driveways, houses, and streets.

The past week had been particularly trying, with no more than a day's break between storms of unusual intensity. As a result, the Russian River was threatening to sweep over River Road, the two-lane artery that connects Rio Nido to Santa Rosa on the east and Guerneville on the west. If the river rose just another 5 feet, Gary and Phyllis well knew, road access to Rio Nido could be cut off for days.

Like most other Californians, Gary and Phyllis had heard about the 1997–98 El Niño, and although they had not paid a great deal of attention to the specifics, they were expecting the winter to be wet. To them, wet meant that the dry gully behind their house would fill with water, earning the nickname Phyllis had given it, White Water. To them, wet meant that the Russian River into which White Water emptied was certain to overflow.

From its headwaters above Ukiah to the small town of Jenner, where it empties into the Pacific Ocean, the Russian River is fed by numerous tributaries. Only a few of them are large enough to bear names, among them Austin Creek, Mark West Creek, Big Sulphur Creek, Dry Creek, Fife Creek, and Cazadero Creek. In reality, water flows into the Russian River the way blood flows into veins, through myriad capillaries that drain a 1,450-square-mile basin. Many of these capillaries, like White Water, are what locals refer to as "winter creeks," because they run with water only during the rainy season.

For as long as anyone can remember, the Russian River has defined the rhythms of life in Rio Nido, a shady hideaway whose name in Spanish means River Nest. During much of the year, the water in the river is low enough to expose wide, gravelly beaches where families gather to picnic and swim. Under the hot summer sun children gather blackberries turned dark and sweet, and adults angle for bass and the occasional trout in deep pools shaded by the rampant growth of willows and alders. Osprey and eagles are often sighted soaring overhead.

During the winter rainy season, however, the Russian River turns into a sullen, surging torrent. First the beaches disappear, then the willows and alders, then entire stretches of River Road. In Rio Nido, and nearby Guerneville as well, some low-lying buildings bear watermarks that extend right up to the rooflines. In that respect, the LaCombes were lucky. Their house was situated in Upper Canyon Three, well above the river's flood zone.

Rio Nido is not exactly a town, just a whimsical hodgepodge of houses that seem to have sprouted, like mushrooms, from the forest floor. Ferns grow in pots along mossy railings. A glittering vest of abalone shells encircles a thick-waisted tree. In places, creeks run under porches, and redwoods seem to grow up through houses. Underfoot is a soft padding of russet needles that muffles noise; overhead arches a deep green canopy that diffuses sunlight. The scent of bay permeates everything.

Rio Nido sauntered into life in the 1920s as a resort community, famous for its lively dances. Over the years, a few dozen San Francisco craftsmen and their families built simple cabins there, which they used mainly as summer and weekend retreats. Then, in the 1960s, Rio Nido, like many of California's backwoods haunts, started attracting a free-spirited population of artists and hippies, who lived there year-round. A bit later, other sorts of people, older people, with steady jobs in nearby towns, also began moving in.

The area had been extensively logged in the nineteenth century, so that save for a few protected stands, all of the trees in the area were second growth. Guerneville, in fact, had once been known as "Stump Town." One of the giant trees processed by its sawmills measured more than 300 feet tall and 70 feet in circumference. In very short order, it was converted into six hundred thousand roofing shingles. Had this same tree been sawn into lumber instead, observed a contemporary, "it would have easily made fifteen thousand feet . . . sufficient to fence with a five-board fence a quarter section of land, to have built a two-story house of ten rooms upon it and a barn large enough to hold one hundred tons of hay and afford accommodations for ten horses, and there would have been enough firewood left to last a family two years."

Gary and Phyllis had bought their house in 1987, around the time they married. It was his first marriage, her second. Both of them were well aware that River Road had flooded the year before, so they weren't wild about the idea of living in Rio Nido. But then their real-estate agent told them about a house that was on the market for $79,000. "It's a pretty house," the agent said. Of course, that's not what Phyllis thought when she saw it. Renters had been living there for months, and as far as she could see, they'd trashed the place. "You can find a filthy, dirty house like that in New York City anytime," she objected in a whisper. "This is California!"

For Phyllis, Gary knew, that was the ultimate insult; she had grown up in New York City but had left, with no desire to live there again. "But hon," he protested, "this house is beautiful!" With the help of Phyllis's daughter, Michelle, Gary won the argument. And it was a beautiful house: two stories, skylights, floor-to-ceiling windows, a spacious kitchen, even a mudroom for yanking off boots. Best of all, it was surrounded by a little clearing, so that the sun poured in on cold winter days.

After they moved in, Gary lavished attention on the garden. He planted flowering trees and roses as well as tulips. He built a concrete patio and a wooden gazebo. He even came up with a scheme to tame White Water, channeling it through the property so that it cascaded down a series of concrete terraces, then rushed through a culvert to join the much larger creek that ran in front of the houses across the street. White Water traced its source to the top of the steep-sided canyon behind the LaCombes' house, and after a big rain, a tree would occasionally tumble over high above and create a logjam. Water, filled with rocks and mud, would back up behind a jumble of branches until it broke through or spilled over, and so all that first week of February, Gary and Phyllis kept a close eye on White Water.

On Monday, they recalled, a huge storm system pummeled virtually the entire coast. In places it dumped more than 12 inches of rain. In Los Angeles, roads leading to the airport flooded, delaying flights even after the worst of the weather passed through. In San Fernando, 78-mile-per-hour winds lifted the roof off a public-housing complex. In Port Hueneme, waves swallowed the last 200 feet of a long fishing pier. In Santa Barbara, gracious palm-lined streets turned into obstacle courses as winds approaching 80 miles per hour bowed, then upended scores of trees, some twenty of which toppled onto cars.

Much farther north, in Santa Cruz County, the Pajaro River crested at its highest level ever. As leaks started appearing in levees, officials hastily ordered the evacuation of ten thousand people who lived within the broad floodplain. Closer to Rio Nido, Fife Creek spilled over its banks, submerging an RV campground in neighboring Guerneville. Driving back from Santa Rosa that night, Phyllis and Gary saw tree limbs down everywhere. "It looks like a war zone," exclaimed Phyllis.

On Wednesday, the sun briefly reappeared. On Thursday, another big storm began to roll in from the ocean. That night, Gary was worried enough to call his boss at the warehouse where he worked to announce that he wasn't coming to work Friday if the river started to flood. "But he made a big fuss, so I said, okay, I'd go in."

The next morning, the storm swept inland, bringing with it battalions of clouds the color of gunmetal and rain that rose and fell in crescendos whenever a squall line moved through. For the most part, though, it didn't seem to be raining all that hard.

Gary went to work and spent eight hours forklifting merchandise on and off the loading dock. He got home about five-thirty in the evening, shortly after Phyllis returned from her own job as a bookkeeper for River Rugs. They had spaghetti for dinner. Then Gary went out to walk the property. The rain was letting up. White Water was running. The culvert seemed clear. All was as it should be, it seemed, except that a tree was down. It would be good for firewood, he thought.

The LaCombes went to bed early. To Phyllis, it seemed like a magical night. "The house had such a warm, beautiful, loving glow," she recalled. "I remember going upstairs and thinking it's so nice to be home. It's so nice to be safe." And they were safe, both Phyllis and Gary thought, well above the reach of the river. They had shelter; they had food; they had each other. What did it matter if River Road should flood? Gary soon fell deeply asleep. Phyllis watched the movie *Ghost* until she, too, began to doze off. Then, around nine-thirty, they were both startled awake. Mai Lin Schultz, the sixteen-year-old daughter of their neighbor Hee-Ran Schultz, had spotted mud flowing down the LaCombes' drive and was frantically pounding on their front door. Still groggy, Gary and Phyllis pulled on their boots and went outside.

At first, Gary thought that it was just the culvert under the driveway that was plugged. While he and the neighbors worked to clear it, Phyllis walked out into the yard, up the garden steps, past the little gazebo, where she and Gary were planning to celebrate their tenth wedding anniversary, and looked up at the steep canyon wall. The narrow gully down which White Water had been flowing, she could see, was now completely plugged. "It's all filled in," she shouted to Gary. "We have a lot of work ahead of us."

Phyllis had gotten about halfway back to the house when a low rumble announced a small landslide. It didn't seem serious, just a shallow slump of rocks and mud that petered out around the gazebo. A bit later on, a couple of firemen stopped by to check things out, concluded that little damage had been done, and then left. As portent piled upon portent, the creek in front of Hee-ran's house suddenly stopped running. That was the creek into which White Water emptied, and the interruption of its flow seemed odd. "I wouldn't go back in that house if I were you," warned a man who lived up the road. "Are you kidding?" asked Phyllis with a perplexed smile.

Gary knew that the plugging of White Water was a bad sign. He worried about a logjam. He worried about a sudden surge of water and mud slamming into the sliding-glass doors that opened from the living room onto the terrace. As reinforcement, he grabbed some plywood from the garage and constructed an impromptu shield. Then, around midnight, he and Phyllis went back inside.

THE LACOMBES were sitting at their kitchen table when they heard what Gary remembers as a "rumbling, snapping, cracking, limb-breaking, tree-crashing avalanche."

Phyllis gasped, "Oh, my God"; then she heard Gary scream: "Ruuunnn!"

She headed for the front door, and heard Gary scream again. "Not that way—upstairs!"

Out the window Gary could see a tidal wave of mud and debris advancing across the patio. It looked to be at least 40 feet tall. He tried to run, but his feet seemed locked to the floor. Phyllis got partway upstairs, then turned to see the root ball of a huge Douglas fir, perhaps 100 feet tall and 15 feet across, hurtling toward the house. She stood there mesmerized. "I thought to myself, *If I'm going to die, I'm going to watch it take me.*"

But at the last minute the tree veered off. At first, it seemed to take aim at the house across the street, where Mai Lin Schultz and her mother, Hee-Ran, stood on the front porch, watching with the horrified fascination of a pair of mice staring into the eyes of a rattlesnake. Then the tree swerved again, sliding into the house next to the LaCombes' with a loud crack.

Over the next few hours the tree slowly pushed the house off its foundation, pinned it against another Douglas fir that was still standing, and split it down the middle like a log. The contents of the house spilled into the muddy torrent that was coursing downhill: cherished photographs, an elaborately painted Hopi doll, a hand-carved sign bearing the name of the family who lived there. Pat Kelly, a volunteer fireman who lived down the street, saw the sign floating by, plucked it from the ooze, and cleaned it off. THE KUDROFFS, it read.

For Phyllis, time seemed to pass in disconnected snatches, like a dream or, rather, a nightmare. "My mind shut down," she said. "All I wanted to do was get the hell out."

In fact, she was near hysteria. "We're going to die," she told the woman who answered her call to 911. "I grabbed her by the shoulders," Gary remembered, "and I told her, 'Hon, you've got to pack.'" But Phyllis stood there staring at him. "I can't even pack for a vacation," she wailed, "and now you expect me to pack for the rest of my life?"

Gary walked into the closet, pulled out a big suitcase, and swept Phyllis's clothes, hangers and all, into it. "I never wear those," she objected, and so he dumped all the clothes on the floor. "Okay," he said. "You pack."

Faced with the overwhelming, the human mind seems to saturate; people do things that later on seem to make little sense and then wonder at them. A chaotic checklist of things to do suddenly rushed into Phyllis's head. In the urgent desire to rescue something useful, she scooped up bottles of vitamin pills. Later on, she and Gary would discover that they'd both forgotten to pack shoes.

When Gary and Phyllis left that night, slogging in their boots through a knee-deep mess of mud, their house was still on its foundation, and their three vehicles—a Bronco, a Dodge pickup truck, a sporty Honda CRX—were still parked in the driveway. But the heavy mass of mud and debris pressed inexorably downhill, and twenty minutes later, as neighbors watched, the house creaked and groaned, then suddenly lurched forward and spun around, like Dorothy's house in *The Wizard of Oz*. A big gash opened at the corner of the kitchen. One after another, the Bronco, the Dodge pickup, and the Honda CRX all slid into the mud.

Hee-Ran and Mai Lin Schultz stood there watching until a big tree nearly slid into the porch where they were standing. Then they followed fire chief Steve Baxman to safety, taking a path that wound along the back of the house to avoid the slurry of mocha-colored muck that, like a spilled milk shake, was streaming down the street out front. Hee-Ran would never forget how eerily beautiful everything seemed. The storm had passed, and the moon had risen. In its ghostly light, giant trees cast giant shadows, and raindrops glistened like tiny sequins.

JIM KUDROFF, the LaCombes' next-door neighbor, learned of the trouble a few hours later. Phyllis, in fact, called him in San Francisco shortly after sunrise. Because of the storm, he, his wife, Paula, and their two young children had not driven back to Rio Nido on Friday; they had stayed overnight in the city instead.

"It looks pretty bad," Phyllis told him. "You should get here as soon as you can."

Jim was worried about the river. How high was it? Was the road still open? It was open, Phyllis said, and so Jim urgently drove from San Francisco to Rio Nido. Taking Highway 101 north, he crossed the Golden Gate Bridge, sped through Marin County, then finally reached Santa Rosa, the capital of Sonoma County's picturesque wine country.

As soon as he turned off the highway, onto River Road, he began to see the familiar, flooded vineyards, bare of leaves and mulched in a bank of fog. Less than a mile before the turnoff to Upper Canyon Three, he passed the brick château where Korbel conducts wine and champagne tastings. Had he stopped there and looked up, he would have seen a big gash in the forested hillside above Rio Nido, a kind of geological graffiti left by the landslide that had just claimed his home.

But Jim was in no mood for stopping; he just wanted to make it to Rio Nido before River Road became impassable. When he finally got there and saw what had happened, he was stunned. In his mind he tried to visualize his house as it had been. Then he tried to cut and paste that mental blueprint onto the shambles he saw before him. Somewhere in the wreckage was a child's bed that he'd built with his own hands. Gaily printed sofa cushions lay exposed to the sky.

Later on, passersby would behold the contents of the house and think of them as strangely exquisite objects, the sort of objects Peter Carl Fabergé might have nested inside a porcelain egg. But the egg that had been Jim's house had been cracked open with deadly force, and so it was that, above all else, Jim at that particular moment felt indescribably grateful. He was grateful that he and his family had decided not to drive up to Rio Nido on Friday night, that the heavy rain had persuaded them to stay over in San Francisco, that no one had been in the house when nature destroyed it. He was grateful that they all still had a place to stay.

For a while, Jim scanned through the rubble, thinking about all the little things that constitute a household's irreplaceable treasures, like the flying bat he'd carved out of redwood for his daughter, Kachina. But he could see nothing. On the way back to San Francisco, Jim called Paula on his car phone.

"It's gone," he said simply.

"What's gone? What do you mean gone?" Paula asked.

"I mean gone," Jim replied. "It's all gone."

WHAT CAUSED THE Rio Nido landslide was not a single fluke storm but a fluky season of storms, the worst winter-storm season California had experienced in fifteen years. The storms started in December. Around the first of the year, there was a deceptive lull. But the lull didn't last, and during January, northern California recorded twenty days of rain. Then, during the first week of February, the immense cyclonic systems rumbling across the North Pacific powered up to full bore.

This was the week of the Rio Nido landslide.

This was the week when students at Stanford University paddled across their flooded campus in kayaks and canoes, when three hundred thousand people were left without electrical power, when a sixty-two-year-old newspaper publisher lost his footing on a narrow bridge and was swept away by raging floodwaters.

This was the week when big chunks of the scenic highway that runs between Carmel and San Simeon tumbled down sheer-sided embankments, cutting off access to Big Sur until May.

This was the week when a forty-six-year-old man in Loma Mar,

California, died after an enormous tree rode a landslide into the bedroom of the house where he lived.

This was also the week when the California storms produced eerie echoes in Florida, where a massive disturbance barreled out of the Gulf of Mexico and exploded like a bomb across the southern tip of the state. That storm has now been inscribed in the pages of popular memory as the Groundhog Day storm, and it is certain that under its baleful skies, no drowsy rodent ventured from its burrow, much less looked for its shadow.

The winds unleashed in the black hours of early morning overturned three hundred airplanes parked at regional airports and whipped up swells that swamped a beamy houseboat anchored at a sheltered harbor. In the middle of the Gulf of Mexico, a crew of swordfishermen had to be rescued by coast-guard helicopters as their boat foundered in 18-foot seas. North of Key West another swordfisherman was crushed by his heaving 51-foot boat as he struggled to secure it to a wooden dock.

Meteorological bookends, the storms that hit California and Florida that first week of February bracketed opposite sides of an enormous continent. Simultaneous in time, yet separated in space, they occurred along coastlines that had different geological histories and weather patterns influenced by different oceans. Yet these storms were tightly linked. The Florida Groundhog Day storm traced its origins to the big cyclonic system that had smashed into California some three days earlier, then sped across the Southwest, nearly splintering apart as it was forced up and over the Rocky Mountains. A day or so later, however, the system regathered its strength, sucking sustenance from the warm, moist air rising off the Gulf of Mexico.

What was amazing, though, was not that the storms that hit California and Florida were linked to each other but that they were both linked to a distant upheaval that, in May 1997, radiated out from the dark blue waters of the tropical Pacific. Over an astonishingly short period of time this climatological convulsion had managed to rearrange the flow of weather across a large portion of the globe, from Canada to Chile, from Indonesia to

Iowa, from Alaska to Australia. There were even hints that the 1997–98 El Niño reached clear across the Atlantic Ocean to affect Europe's weather as well.

For months on end, readers of newspapers and watchers of TV were regaled with accounts of strange and wondrous happenings. Chileans, they learned, basked on strangely balmy beaches in the middle of their winter, while, some six months later, Mexicans shivered under a chilly coverlet of fresh-fallen snow during theirs. Peruvians wondered at a great lake that appeared in the middle of the Sechura Desert, while Indonesians suffered through a scalding drought that shriveled crops and allowed wildfires to blaze out of control.

As the images of fires, floods, tornadoes, and snowstorms flashed across the world, the troubles of middle-class Americans like the LaCombes seemed to merge with those of hungry subsistence farmers on the islands of New Guinea and Borneo, flood-ravaged villagers in Kenya and Somalia, and landslide victims in impoverished areas of Ecuador and Peru. And amazing as it seemed, these global reversals of fortune all shared the same cause—El Niño, the warm side of a climate cycle that has only recently been recognized as the most important influence on monthly weather patterns apart from the seasons.

Climatologists have a term for the meteorological sleights of hand by which El Niño turns dry places wet, wet places dry, cold places warm, and warm places cold. They call them teleconnections, literally long-distance links, and by the end of the twentieth century these links had begun to seem so strong, so obvious, that it was hard to remember that people everywhere had not always known about El Niño, that only toward the end of the twentieth century had they keyed into its power to reconfigure the world.

IMPRINT OF A MIGHTY FIN _____

Long before they became consciously aware of it and began to understand the physics that caused it, people around the world had marched to El Niño's syncopated beat. A surprisingly detailed record of El Niño's comings and goings, for example, has emerged from ancient Egypt, where, for thousands of years, the rhythm of life revolved around the annual ebb and flow of the Nile.

The whole system of agriculture that sustained the pharaohs and their subjects, it is well known, depended on the floods that began in the southern part of the Sudan in April, then spread north, to Cairo, cresting in September or October. At their height, the floodwaters of the Nile covered the alluvial plain to depths of 3 or more feet.

The floods were not only important as suppliers of water in this desert land but also as a source of silt that gradually built up to form a broad and fertile planting medium. So much soil was carried by the floods that the ancient Egyptians referred to the Nile as the Ar, the Black River. They also referred to the land in terms of its colors: the Red Land of the desert, the Black Land of the river. The Black Land, we now know, was a gift from the highlands of Ethiopia, and over the course of thousands of years, it accumulated to depths of 20 feet and more.

It was the rainfall in Ethiopia that was so critical to Egypt, and it was this rainfall, scientists later established, that El Niño so powerfully affected. To understand why, one needs to understand that the Nile has two main tributaries. The White Nile traces its source to a large drainage basin that empties into the blue waters of Lake Victoria. The Blue Nile originates in the Ethiopian highlands; it is fed by streams and rivers that run full during the main rainy season, which is known as *kremt*. Another important tributary of the Nile, the Atbara River, also traces its source to the Ethiopian highlands.

When the *kremt* rains are generous, the Blue Nile surges into the White Nile with boggling force; the flow at Khartoum sometimes exceeds two hundred and fifty thousand cubic feet of water per second. At flood stage, the Blue Nile is so powerful that it actually blocks the White Nile from flowing north. But when the *kremt* rains fail, the Blue Nile slows to a trickle. For many millennia, then, the success and failure of the Ethopian rains determined the fate of Egyptian agriculture. And the rains did sometimes fail.

Consider, for example, the famous Famine Stele that, at the end of the nineteenth century, was discovered at Sehel, one of the islands clustered around the first cataract of the Nile. The stele itself is not quite two thousand years old, but there is reason to think of it as the scrawled-in-stone equivalent of a late-edition book. That is, the stele's hieroglyphic text purports to encode a document that is 4,750 years old, and among the inscriptions is the lament of a ruler whose people have been seized by the eagle talons of hunger.

> I am mourning on my high throne for the vast misfortune, because the Nile flood in my time has not come for seven years. Light is the grain; there is lack of crops and of all kinds of food. Each man has become a thief to his neighbor. They desire to hasten and cannot walk. The child cries, the youth creeps along, and the old man; their souls are bowed down, their legs are bent together and drag along the ground, and their hands rest in their bosoms. . . . Torn open are the chests of provisions, but instead of contents, there is air. Everything is exhausted.

National Geographic, July, 1917

The ruler was supposedly the Third Dynasty pharaoh Zoser, builder of Egypt's very first pyramid, the Step Pyramid at Saqqara. Zoser is also said to have built a temple on the island of Elephantine to honor the god Khnum, "the lord of barley and wheat, fruit and flowers, birds and fish and animals." It was Khnum, sitting on his throne at Elephantine, who supposedly controlled the flow of the river; he did this, it was said, by opening and closing the gates of vast subterranean caverns that held the source waters of the Nile.

To keep Khnum happy, Zoser apparently decreed that his people should offer up an annual tithe in his honor, amounting to one-tenth of their harvest, their fish catch, their game animals, and their goods of trade, and it was said that, after that, the annual floods returned, bestowing on the thirsting land their priceless blessing of water. Interpretation of the inscription was tricky. One couldn't tell whether one was reading a true account of a famine that had actually happened or, more likely, an embellished account laden with political spin designed to suit the purposes of Zoser and his successors.

Of slightly more recent vintage was the seven-year famine recorded in the biblical account of Joseph and the pharaoh's two dreams. In those dreams, which are described in Genesis, seven fat cows gave way to seven thin ones; seven plump ears of grain, to seven scrawny ones—and then, to the pharaoh's horror, the scrawny cows and ears of grain devoured the plump ones. It was Joseph, of course, who correctly interpreted the dreams and arranged for food to be stored during the seven years of plenty. Thus, when the seven years of want arrived and famine stalked the world, "there was bread in all the land of Egypt." The date of this famine would appear to have been around 1700 or so B.C.

The ancient Egyptians did not know about the importance of the rains in Ethiopia, nor did they realize that El Niño suppressed them. But beginning around five thousand years ago, they judiciously monitored water levels in the Nile, taking regular measurements with yardsticklike gauges that came to be known as Nilometers. Indeed, it is said that Joseph himself installed a Nilometer at Bedreshen on the west bank of the Nile.

The first Nilometers were just rows of slashes carved on rock or stone walls; eventually they came to be supplemented by riverside staircases and wells with marked cisterns. According to an Arab chronicler and physician named Abd al-Latif, a rise of more than 17 cubits (around 52 feet) was an occasion for joy, ensuring a harvest that would produce a two-year supply of food. Less than 16 cubits required active watering of the fields, which, even so, produced lower yields.

In the mid-1980s, using these observations as a baseline, Oregon State University oceanographer William Quinn, a pioneering investigator of El Niño's past, established a scoring system that rated the flood failures on a scale of one to five, five being the most severe. Then he reviewed more than one thousand years of Nilometer records. Since A.D. 622, he deduced, there had been twenty-two major failures of the Nile flood, and those that occurred in the nineteenth and twentieth centuries were clearly associated with El Niño.

In the early Nilometer record, some of the years that stood out as particularly dire were A.D. 1096, 1144, and 1200; Quinn singled out the last as the greatest Nile flood failure in the record. That was the year when, according to the chronicler Latif, whole villages turned into ghost towns and butchers did a brisk business in human flesh. At the time, Quinn speculated that the A.D. 1200 flood failure and famine might have been associated with the same El Niño that, on the other side of the world, took out the food-production system that sustained the imperial city of Chan Chan.

LONG AGO the Spanish recorded a "legend" of a more northerly Chimu people who were once early rivals of Chan Chan. They were a powerful nation until their king incurred the gods' wrath, and as punishment it rained for thirty days and nights. Devastating flooding ceased only when the populace rose up, bound the king hand and foot, and threw him into the ocean. There followed great famine and pestilence, lasting for countless years, and then Chan Chan's armies swept across the land, conquering all.

Fred Nials, Eric E. Deeds, Michael E. Moseley, Sheila G. Pozorski, Thomas G. Pozorski, and Robert Feldman, *Bulletin of the Field Museum,* 1979.

The year was 1972, the place was the Rio Moche Valley in northern Peru, not far from the colonial city of Trujillo, and participants in an archaeological project sponsored by Chicago's Field Museum of Natural History were roaming over rocky terrain, spiderwebbed with dry riverbeds and gullies, when quite suddenly "it began to rain," as if "nature had lost equilibrium!" members of the expedition exclaimed in an article written for the *Bulletin of the Field Museum*.

For days, they continued, "erratic showers soaked, then saturated the normally parched landscape. With no vegetation to hold it back, runoff transformed dry washes and quebrada (normally dry ravine) channels into torrential streams. Flash floods cut all the northern highways, stranding not only the region we were in but the rest of the nation as well. Settlements were flooded, adobe houses collapsed, and mud choked down the life-sustaining canal systems."

Such rains would have seemed heavy virtually anywhere in the world. According to a rain gauge positioned at a nearby fishing village, some 200 millimeters, nearly 8 inches, fell in that one place in just three days. But the wonder of *these* rains was that they fell on Peru's coastal desert, a narrow thong of rock and sand that counts as one of the planet's most arid places. In a normal year, precipitation in these parts rarely exceeds 55 millimeters, or little more than 2 inches, and in many areas, it is less than 2 millimeters, barely a moist whisper.

By this time, of course, scientists had begun to key in to the power of El Niño, and so there was no doubt about the cause of the trouble. But the experience of being caught in the middle of such an event caused the archaeologists to wonder how often El Niño had unleashed similarly devastating rains in the past.

And so they began to investigate.

The flooding that occurred in 1972, the archaeologists established, was nothing to that unleashed in 1925 when the Rio Moche rose so high that it took out the railroad bridge that linked the city of Trujillo to the capital and threatened the ceremonial center of an ancient empire. In that year, the archaeologists went on to document, spillover from the rising river reached partway up the base of the Huaca del Sol (Temple of the Sun) and the Huaca de la Luna (Temple of the Moon), massive structures built of hundreds of thousands of adobe bricks.

What struck the archaeologists as so unusual about the 1925 El Niño, however, was not so much the flooding of the Rio Moche itself, as dramatic as it was. It was rather that, for the first time in living memory, the Rio Seco to the north of the Moche had gushed with water. Once upon a time, the archaeologists knew, this mere ghost of a quebrada, as elaborately braided as a pigtail, had been an integral part of the irrigation system that had been built in the vicinity of Chan Chan, seat of the Chimor Empire.

The Chimu, as the people who built Chan Chan are called, flourished from around the tenth to the fifteenth centuries. Like their predecessors the Moche, they were able to settle and thrive in this seemingly inhospitable desert environment because of two very important natural assets. First, their proximity to the coast put them adjacent to one of the richest fisheries in the world; second, their proximity to the Andes placed them near rivers that were seasonally replenished by runoff from mountain snow and rain.

To understand this, one has only to visit the ruins of Chan Chan. Even today, from inside the winding corridors excavated by archaeologists, one can hear the distant pounding of surf; from the top of a viewing platform, one can see the green fields that modern farmers have cadged from the desert.

The irrigation system the Chimu built was extremely sophisticated; it channeled water into an intricate network of canals, which, in turn, fed into a series of walled fields. The purpose of the canals, of course, was to divert water for agriculture before it drained into the sea. Not surprisingly, as the population of Chan Chan grew, so did the demand for water. At one point, the Chimu undertook the construction of a canal more than 40 miles long that tapped into the Rio Chicama in an adjacent river valley.

This intervalley canal, it has been suggested, fell afoul of the tremendous forces that, even today, are pushing the Andes up. Since 500 B.C., the Field Museum's archaeological field team determined, the area around Chan Chan had risen by more than 20 feet. But tectonic uplift was not the only source of serious disruption. At some point, the archaeologists found, the sophisticated canal system around Chan Chan had been utterly devastated, and they unearthed a chain of circumstantial evidence that pointed to a massive flood—two to four times larger than the flood that occurred in 1925—as the wrecker's ball. The surging waters of the Rio Seco, it appeared, had overwhelmed the canal system, eroding its walls and filling it with sediment.

Conveniently, the floods deposited fine-grain layers of silt and sand that pretty well tracked the water levels. Flood deposits in the ruins of an ancient village in the vicinity, for example, indicated that water from the river had likely rushed in while houses were still in use, no doubt sweeping many to their deaths. Moreover, this did not seem to be just a local flood along the Rio Seco. There was dramatic evidence that the Moche, too, had surged over its banks. At the Huaca del Sol, the waters reached as high as 25 feet above the present level of the river, and at the Huaca de la Luna, two times higher. Indeed, the flood along the Moche probably inundated an area more than 2 miles wide, with water nearly 50 feet deep!

In the years that followed the initial investigation, the story of Chan Chan's flood became considerably more complicated. One faction of archaeologists now places this flood along the Rio Seco at the beginning of the twelfth century, around A.D. 1100, while another has fast-forwarded the date to the early fourteenth century. But more important than the dispute, perhaps, is that both sides do agree about one thing: The floods did not spell the end of the Chimor Empire.

Quite the contrary. As a solution to the socioeconomic crisis the floods generated, the rulers of Chan Chan appear to have been inspired to expand their food supply through conquest, thereby establishing dominion over a 600-mile-long strip of coast that stretched from the vicinity of present-day Lima to the Gulf of

Guayaquil. It is a curious footnote that among the areas that Chan Chan's armies overwhelmed was the rich agricultural country of the Lambayeque Valley, the very place where, legend has it, a long-suffering people rebelled against their ruler after enduring thirty days and nights of nonstop rain.

For this reason, some archaeologists have speculated that the legend might be substantially true, in which case the name of the king would have been Fempellec, the last ruler of the Naymlap dynasty. Fempellec supposedly brought disaster upon his people after falling under an evil spell. At the urging of a sorceress, it was said, he removed from its place of honor a green stone idol that for centuries had protected his kingdom from harm.

Over time, Chan Chan grew into a large imperial city. Its external walls enclosed an area of nearly 8 square miles, and at its height, it may well have boasted some forty thousand residents. The lower classes are believed to have occupied houses and workshops built of cane, while the nobility lived and worked in substantial adobe brick buildings. Chan Chan's kings were buried, along with their attendants, in the same sumptuous palaces they occupied during their lifetimes. In an article written for *National Geographic,* archaeologist Michael Moseley and a colleague described excavating these tombs and finding "the skeletal remains of young women stacked up like cordwood."

In 1470, when the Inca conquered Chan Chan, they were as agape at its splendors as any of the barbarians who rode into Rome. Bedazzled, they beheld earrings decorated with images of reed boats and fish, staffs topped by long-necked cormorants, ceramic pots decorated with the figures of fishermen, doorways encrusted with silver. They looted, then left, bringing back to Cuzco, their own imperial city, not just a wealth of gold and silver and elaborately woven textiles but also the artisans who made them.

A COMMON THREAD _____

Of all the glimpses into the dimly lit corridors of El Niño's past, none seem more tantalizing or more elusive than those offered by the poetry, proverbs, and stories that celebrate the Indian monsoon, especially the summer monsoon, the subcontinent's main season of rain. In the sixteenth century, for example, the poet Guru Nanak penned the following lines:

> Oh my heart, rejoice! It's Sawan
> The season of nimbus clouds and rain

Almost a thousand years earlier, another bard had written:

The rainy season had arrived. Rivers overflowed their banks. Peacocks danced at eventide. The rain quelled the expanse of dust as a great ascetic quells the tide of passion. The chataka birds were happy. Lightning shone like a bejeweled boat of love in the pleasure pool of the sky. . . .

It was, of course, the always haunting possibility that the rains would fail that made their arrival such an occasion for joy. For when the oppressive heat of April and May lingered unbroken into June and July, then famine, pestilence, and death stalked those who lived under pitiless skies that were armed "with a burning sword," wrote the English poet Rudyard Kipling.

The reign of Emperor Shah Jahan, the seventeenth-century ruler who built the hauntingly beautiful Taj Mahal, was marred by a horrific famine in which, according to one account, "the number of the dead exceeded all computation or estimate. The towns and their environs and the country were strewn with human skulls and bones. . . . Human bodies, dried in the sun, were steeped in water and devoured by those who found them."

In the Indian subcontinent, British historian David Arnold has observed, virtually every century would appear to have been marked by one or more great famines, to the extent that peasants used them as a handy ruler for marking off time. They would say that so-and-so had been born or had gotten married or had died in the year of a certain famine, or so many years before it or after it ended. Some of the worst famines, according to Arnold, were "named after vengeful deities or were seen as marking the onset of the Kali Yuga, the Hindu age of suffering, corruption, and human misery. One particularly widespread and destructive Indian famine in the fourteenth century was recalled for generations afterward as Durgadevi, after the awesome Hindu goddess of this name."

But while all centuries were marked by famines, none were better chronicled than those that struck after the British arrived in the early part of the eighteenth century. In 1770, for example, up to one-third of the people of Bengal were said to have died for want of food; between 1789 and 1792, famine is thought to have claimed one-half the population of Madras. And then, not quite a hundred years later, came the famine that would launch the scientific search for the large-scale climatological upheaval that became El Niño as we understand it today.

THE STATE OF things [grew] . . . immeasurably worse . . . no camp of three thousand rose morning after morning without leaving thirty of its number upon the ground to rise no more. In the interior the distress was dreadful. One gentleman passing down a ghaut in the Wynaad in July counted twenty-nine dead bodies on the road; a coffee planter, seeking shelter from rain in a hut, found six decomposed bodies in it. People died of starvation in the streets of Madras. . . . One word of tribute must be paid to the law-abiding tendencies of the people in general. In spite of their terrible distress, notwithstanding the claims of their empty bellies, there were remarkably few robberies.

Henry Digby, honorary secretary, the India Famine Relief Fund, *The Famine Campaign in Southern India 1876–1878*

The seeds of the trouble chronicled by Digby were sown in 1876, when neither the summer nor the winter monsoon brought more than scant smatterings of rain. "The country is an absolute desert," a Bombay official noted in his district report for August of that year. "Not a blade of grass nor a particle of grain is to be seen anywhere. The rivers and wells are drying up, and the cattle are dying of absolute starvation. The people are looking very thin and ill, and though there have been no deaths from starvation, it is very evident that something must be done speedily to alleviate the distress."

As the shadow of approaching scarcity lengthened, merchants started hoarding supplies of grain; overnight, it seemed, the price of rice and wheat rocketed beyond what many could afford. In the district of Madras, tens of thousands had already perished by May 1877, when a caterwauling cyclone dumped more than 20 inches of rain in the course of three days. The deluge brought more misery than relief, however, since most of the badly needed moisture simply ran off uselessly into rivers and streams.

Exponentially, it seemed, the desperation of the famished hordes increased. "The ties of nature were ruthlessly severed," lamented Digby. "On any and every day mothers might be seen in the streets of Madras offering children for sale, while the foundling section of the Monegar Choultry was full of infants found by the police on the roads, deserted by their parents, some with the marks of recent birth still upon them."

During June 1877, Digby reported, hope briefly rose as some communities reported receiving "heavy 'plumps' of rain," but the coverage proved to be extremely spotty. As the summer sun bore down with unbridled force, once-verdant fields grew to resemble a wasteland, "withered and bare and desolate." Around Bombay, nature also taunted desperate villagers with welcome showers, prompting a hopeful rush to the fields. But "after the first heavy dashes of rain," Digby observed, "the skies once more became as brass, and for nearly two months no rain fell."

By the time the rains resumed, Digby estimated, some five million people may have died, at least 1.5 million in the region around

Madras alone. The majority, it should be noted, did not die directly of starvation, but rather, weakened by lack of food, succumbed to diseases like malaria and cholera. And the raw numbers, appalling as they were, did not begin to reflect the staggering scale of the tragedy. In 1878 the population in the hardest-hit areas declined 25 percent from what it had been in 1871 and 32 percent from what census takers estimated it should have been. That's because, until famine inverted the demographic curve, these areas had registered an annual excess of births over deaths in the neighborhood of 1.25 percent.

The horrors of this particular famine, among the worst in India's long history, did not go unnoticed in England, where appeals for help sang out from the pages of newspapers, the desks of schoolteachers, and the pulpits of churches. Letters, along with money, began to flow to Madras, whose leaders had responded to the crisis by establishing a national famine-relief fund.

One of the letters came from Florence Nightingale, who observed, "If English people know what an Indian famine is—worse than a battlefield, worse even than a retreat; and this famine, too, is in its second year—there is not an English man, woman or child who would not give out of their abundance, or out of their economy." She went on to note that she saw the committee's appeal for funds "this morning only," and "thanking god that you have initiated this relief, I hasten to enclose what I can—25 pounds; hoping that I may be able to repeat the mite again; for all will be wanted."

In all, Nightingale and her contemporaries anted up 800,000 pounds sterling in charitable contributions.

BY THAT TIME, barometers and rain gauges were positioned in many locations around the world, so that in addition to contemporaneous eyewitness reports, scientists also had access to real meteorological data. More than a hundred years later, two climatologists, George Kiladis of the University of Colorado and Henry Diaz of the National Atmospheric and Oceanic Administration, mined that data to produce a portrait of the powerful El Niño in which the famine had been embedded. They were inspired to do so by El Niño of 1982–83, the most powerful El Niño of the twentieth century until El Niño of 1997–98 came along.

Coastal Peru, they established, was extremely wet in 1877 and 1878. By contrast, Jakarta, the capital of Indonesia, was extremely dry. In fact, it received less than a third of its normal allotment of rain. Dry conditions prevailed not only in Australia and southern Africa but also in northern Brazil, where a severe famine stalked the northeast, sending tens of thousands of poor subsistence farmers on a death march to the coast. All told, that famine is thought to have claimed between five hundred thousand and a million lives. And, yes, although Kiladis and Diaz did not mention it in that particular paper, in 1877, the Nile flood likewise came in at very low ebb.

And the United States? The winter of 1877–78, Kiladis and Diaz concluded, was the second wettest San Francisco had experienced in 125 years; by contrast, 1982–83 qualified only as the third wettest. At the same time, St. Paul Island, in the Bering Sea, which normally sees quite a bit of rain, turned extremely dry in 1877–78. The temperature profiles Kiladis and Diaz dug up also fit the pattern. During the winter of 1877–78, not just the northern two-thirds of the continental United States but the entire world appears to have warmed: "1878 was the warmest year globally from 1856 to 1981," they observed.

How, one wonders, could a phenomenon that was so important, so influential, have been overlooked for so long? There were a number of reasons.

For starters, there was the temporal problem. The human psyche picks up most readily on patterns that are highly regular, that repeat every year like the seasons or that lock into place like the ice ages. El Niños, by contrast, come and go on a much more erratic schedule; they also vary in strength, so that their impacts on any one area will veer from pronounced to barely noticeable.

Spatially, too, El Niño challenges the mental capacities of humans, who only in the last few centuries have been able to understand that storms are not born just over the horizon but in the far beyond. In fact, it was precisely because El Niño was so large, so variable, and because it played across such a long fetch of space and time, that its full, monstrous measure was not taken sooner. And as if that were not challenge enough, scientists had to contend with an even greater obstacle, the fact that El Niño represented something unique, an altogether fantastical confabulation of wind and waves.

As the next chapter will show, it took nearly a century before scientists realized that failures of the Indian monsoon, weakenings of the Nile flood, and torrential downpours in Peru and in California were so often connected to the same globe-girdling upheaval that they were among the many disasters to be anticipated when El Niño breached in the faraway waters of the tropical Pacific, a sea, wrote Herman Melville, filled with "gentle awful stirrings."

CHAPTER **TWO**

There Leviathan
Hugest of living creatures, in the deep
Stretched like a promontory sleeps or swims
And seems like a moving land: and at his gills
Draws in; and at his breath spouts out a sea.
JOHN MILTON, *Paradise Lost*

THE CHRISTMAS CURRENT

Framed by an amphitheater of dun-colored rock, the sleepy port city of Paita, Peru, has ever stood with its back to the desert and its face to the ocean. On one side lies desiccated wasteland; on the other, one of the most productive fisheries in the world. On one side, life struggles under the tall rain shadow cast by the Andes; on the other, life flourishes in nutrient-rich water that usually ranges between a low of 50 degrees Fahrenheit and a high of 71. The chill of these waters, unusual for the tropics, was noted in 1802 by the great explorer Baron Alexander von Humboldt, and it is the reason the Peru Current (which begins near Antarctica and tracks north for 3,000 miles) is also known as the Humboldt Current.

Not just around Paita, but all along the coast of Peru, the same startling contrast prevails. Aside from the green oases that border the coastal rivers, the barrenness of the desert is unrelenting, a tedium of brown only occasionally relieved by smatterings of shrubs and stunted groves of wraithlike algaroba trees. The ocean, by contrast, teems with all manner of creatures large and small. In a good year, trillions of the quicksilver fish known as anchoveta swarm through Peruvian waters, so that if anchoveta were stars, it has been said, they would fill up dozens of galaxies the size of our own Milky Way.

In a bygone era, Paita was a place of more than minor importance. After the Spanish conquest, Paita became a regular port of call for merchant ships laden with the spoils of conquest and, thus, a frequent target of raids launched by English buccaneers. In 1579, Paita was attacked by Sir Francis Drake, and in 1741, by Sir George Anson, whose men began by ransacking the town, then went on to burn it down and sink all but one of the merchant ships anchored in its harbor.

Today the fort built to fend off these attacks lies in ruins, and life in Paita has settled into a slow, hypnotic rhythm. In the morning and late in the afternoon, fishing boats of small-to-middling size glide in and out of the gauzy mist that hangs above the water like a scrim. Occasionally, a cargo ship stops by to load up with fish meal processed by local plants or cotton and other crops grown in the irrigated fields along the Piura and Chira Rivers. Otherwise, Paita drowses beneath a patina of seedy melancholy, lulled to sleep by the glare of the sun and the gentle sea breeze that, shortly before noon, starts blowing in from the ocean, cooling the air without bringing rain.

Perhaps it was Paita that Melville had in mind when he described in *Moby Dick* "the tearlessness" of Peru's desert skies. For among the grab bag of historical footnotes that pertain to Paita is that Melville stopped there in 1841 with the purpose of meeting Manuela Sáenz, the former mistress of the great revolutionary hero Simón Bolívar. Exiled from her native Ecuador, Sáenz spent the last twenty-five years of her life in Paita, eking out a living selling cigars. "I admire you, not as the victor crowned by honors, but as the defeated," Melville gallantly said when they met, according to Sáenz's biographer Victor Von Hagen.

But of all Paita's holds on collective memory, the most peculiar, and possibly the most enduring, is the role it is said to have played in naming El Niño. According to folklore, at least, it was sailors from Paita who first noticed the strange warm current that appeared off their shores around Christmastime and who called it La Corriente del Niño, the Christ Child's Current, in honor of the season.

In contrast to the north-flowing Peru Current, these astute seamen noticed, the warm El Niño current coursed down from the equator. Moreover, it appeared every year, keyed to a seasonal shift in wind patterns, and it usually came in December and went in March without causing a big ado. Occasionally, however, La Corriente del Niño displayed an unruly side to its personality. At such times, it was intemperately warm. It surged south with churlish force. It appeared early and lingered late. And it coincided with the arrival of that rarest of all commodities in this arid coastal zone: rain.

Then, and only then, lakes formed in the Sechura Desert, quebradas turned into rivers, and rivers widened into inland seas. At such times, flash floods roared through desert cities and towns. An American living in Talara, just north of Paita, described the effects of the 1891 El Niño in a letter to a friend:

A low-hanging, densely clouded sky, intense heat, great humidity and an oily sea—so it began. The temperature varied little during the twenty-four hours. Everything we possessed became mildewed. The rain fell in inconceivable torrents for weeks. . . . After dark the sea broke in phosphorescent lightnings all along the coast. . . . The desert became a garden. . . . Almost every evening we were pestered by insects of ever-changing variety and of every size and shape imaginable, that dashed themselves against the lamps and had constantly to be brushed off the table. . . . Even more remarkable were the spiders that came into the house, dozens of species we had never seen before and never saw afterward. . . .

Just before the rains the Chira was a goodly stream, unfordable, flowing close to the cliffs on the southern side of the valley. The important town of Amotape was built on a raised shelf on the other northern side. Between the town and the river was a level plain . . . cultivated by irrigation, with farms, gardens, banana orchards and the like, and a particularly beautiful grove of old *algaroba* trees, prized as you know for their bean-like fruit or seed pods which take the place of oats for mules and horses. . . . When I visited Amotape shortly after the rains had ceased, I found that the orchards and gardens and the beautiful old grove of trees had all been swept away. The river in subsiding had cut an entirely new channel for itself and now flowed just below the town. . . . I am sorry to say I do not remember how long the rains lasted.

No doubt because its impacts were so dramatic, the 1891 El Niño appears to have been the first to spark serious scientific interest. Speaking at the Sixth International Geographical Congress, which was held in London in 1895, Señor Federico Alfonso Pezet declared, "As this countercurrent has been noticed on different occasions, and its appearance along the Peruvian coast has been concurrent with heavy rain in latitudes where it seldom, if ever, rains to any great extent, I wish, on the present occasion, to call the attention of the distinguished geographers here to this phenomenon, which exercises, undoubtedly, a very large influence on the climatic conditions of that part of the world."

Thirty years later, in 1925, another powerful El Niño struck. This time a young ornithologist by the name of Robert Cushman Murphy happened to be on hand to record it. The El Niño current, he reported, was so strong that commercial ships bound for Panama fell behind schedule by as much as ten hours over a distance of 800 miles. It was so strong that it brought to the coastal desert of Peru alien exotica from the jungles of Ecuador, among them "live reptiles (that had) . . . floated ashore on rafts of green plants." Amid the debris clogging the harbor of Talara, Murphy happened upon "a venomous serpent locally known as a macanche *(Lachesis lanceolatus)* and an Ameiva-like lizard."

Writing in the *Geographical Review*, Murphy also noted the meteorological impacts associated with the El Niño current, among them the flooding of the Rio Moche in the vicinity of Trujillo, several hundred miles down the coast from Paita. Named after the Spanish city that was Pizarro's birthplace, Trujillo boasts an elegant Plaza de Armas surrounded by sixteenth- and seventeenth-century buildings with beautiful wrought-iron balconies. It is flanked, on the north, by the ruins of the ancient city of Chan Chan and, on the south, by the Huaca del Sol and Huaca de la Luna, ceremonial centers of the even more ancient Mochica civilization.

As Murphy summed up the state of knowledge at the time:

> El Niño, the warm countercurrent, is a well-known phenomenon in northern Peruvian waters, commonly appearing around Christmas time and flowing southward. . . . During a longer cycle, traditionally believed to be seven years . . . its manifestations are more pronounced and extensive. At still-longer intervals, representing in the present instance a rhythm of thirty-four years, El Niño and its associated phenomena attain their maximum expression. The combined effects during the early months of 1925 were far greater than those of any season since 1891 and probably greater than any year of record.

Especially astonishing to Murphy, who went on to head the department of birds at New York's American Museum of Natural History, was the massive die-off of seabirds associated with El Niño of 1925. As Murphy observed, these birds—pelicans, cormorants, boobies—were utterly dependent on the vast schools of anchoveta that teemed offshore, often swimming so close to the surface that they resembled fast-moving rivers of molten silver. The fish were so ridiculously easy to catch, Murphy quipped, that for a hungry seabird, eating was usually about as complicated as snatching sardines from a tin.

In 1925, the animated rivers of anchoveta very suddenly stopped running, and soon afterward Murphy spotted the putrefying carcasses of seabirds bobbing in the ocean and piling up along the shore in malodorous numbers. Harbors and bays were fouled by a rotting organic mass, which released hydrogen sulfide and other noxious gases. In response, the bottoms of fishing boats turned black, giving rise to the nickname that residents of ports like Paita and Callao often use instead of El Niño—El Pintor, or the Painter.

What caused the carnage? At one point, Murphy speculated that the warmth of the El Niño current, in combination with toxins produced by El Pintor, was the main injurious agent. Just as a lengthy heat wave scorches vegetation on land, he speculated, so the fetid breath of the El Niño current "seared" the luxuriant pastures of microscopic phytoplankton that formed the base of the marine food chain. "The Bible tells us, 'All flesh is grass,'" Murphy observed. "This is true of the oceans no less than of the land."

Murphy's colorfully framed hypothesis proved to be a bit off target. The collapse of the phytoplankton at the base of the marine food chain, scientists have gone on to establish, is primarily due to the depressed nutrient content of the water rather than its elevated temperature. But Murphy and his contemporaries were only beginning to grope their way through the labyrinth that would eventually lead to a fuller appreciation of El Niño's complexity, its staggeringly large dimensions. They still thought of El Niño largely in terms of a warm current that came down from the equator; they still thought of El Niño's effects as confined to the vicinity of Peru.

In 1931, a German oceanographer named Gerhard Schott took a tantalizing first step toward constructing a much larger view of El Niño. From observations made by merchant seamen in 1891 and 1925, he deduced that El Niño was not "a local phenomenon and by no means originates in the warm, humid Gulf of Guayaquil. It is a very extensive oceanic disturbance which in turn is caused by an equally vast atmospheric disturbance." By "vast," Schott meant that the disturbance might extend off the South American coast for a thousand or so miles.

In fact, El Niño is so much larger. Its oceanic footprint, from heel to toe, measures nearly 9,000 miles long and extends clear across the tropical Pacific, all the way from Peru and Ecuador to Papua New Guinea and the Indonesian archipelago. Its atmospheric footprint is bigger still; it wraps all the way around the earth. And through the atmosphere El Niño reaches out to cause a sympathetic warming of the tropical Atlantic and Indian Oceans as well.

WHAT IS EL NIÑO? As scientists now understand it, El Niño is one side of a naturally occurring cycle that exercises the largest month-to-month influence on earthly weather patterns after the seasonal march of the earth about the sun. The other side of the cycle, which was not identified until much later, has come to be known as La Niña, the girl child in Spanish.

El Niño's most celebrated signature takes the form of a gigantic pool of warm water that extends from the International Date Line to the South American coast. La Niña, by contrast, brings an influx of unusually cool water into the same region. The difference is that the water to the east of the date line is customarily cool, so that El Niño represents an exceptional state of the ocean; La Niña, an exaggeration of its normal condition.

Like storms, El Niños and La Niñas come in an array of strengths and sizes. Like storms, they can start out strong, then fizzle or perform the opposite trick by pretending to fade away, only to come roaring back. But there the analogy breaks down. Storms are transient displays of nature's power, local or regional phenomena that fall under the rubric of weather. El Niños and La Niñas, by contrast, are planetary-scale weather-makers, master manipulators of the context in which weather takes shape.

No less than the seasonal cycle, the El Niño cycle is driven by the sun, but by the concentration of solar heat in the equatorial region as opposed to the seasonal changes in its distribution.

What happens is this: Over the course of a year, two years, often more, the western basin of the tropical Pacific slowly collects much of the excess heat that the sun so lavishly bestows along the earth's midline bulge. Then, almost like a lake overtopping an invisible dam, this huge reservoir of warm water releases its pent-up energy in one long, sustained burst. That burst is El Niño.

Neither El Niño nor La Niña is a purely oceanic phenomenon. Each can equally well be characterized by striking changes that occur in the atmosphere. El Niños, for example, involve a large-scale weakening of the trade winds that blow from east to west across the equatorial ocean. In La Niña years, the trade winds do not wane, but wax even stronger. These winds got their name because merchant ships plying routes around the world relied on them to fill their sails; indeed, the trades are highly dependable except when El Niño comes to call.

El Niño and La Niña are highly complex phenomena. Among the climatologically important parameters they affect are these: the speed and direction of currents in the ocean; the speed and direction of the winds that drive the currents; the temperature of the water at the surface of the ocean; the pattern of atmospheric pressure at the surface of the ocean; even the slope of the surface of the ocean as it trends from east to west. El Niño's impact on these parameters, though, is greater than La Niña's. Again, this is because El Niño represents true aberration, while La Niña seems more like normal writ large.

Loosely speaking, El Niño plays the summer to La Niña's winter, but with a very important difference. The swings between the seasons, occasioned by the earth's annual march about the sun, are nearly as regular as the ticktocking pendulum of a grandfather clock. By contrast, the swings between El Niño and La Niña resemble the gyrations of a complex pendulum whose coupled arms never retrace precisely the same path.

In this case, the pendulum can be thought of as having two arms of unequal length. The longer arm belongs to the ocean; the shorter arm, to the atmosphere. The difference reflects the fact that the atmosphere responds very quickly to the changes that occur in the underlying ocean—winds, for example, respond to a rise in sea-surface temperatures almost instantaneously—but preserves no lasting memory of the changes it perceives. By contrast, the ocean senses changes in the atmosphere, reacts to them sluggishly, but then remembers and remembers and remembers.

As those who've played around in physics labs know, the oscillations of multiarmed pendulums can spontaneously produce a fascinating display of dynamic behavior. As one looks on, the shorter arms of the pendulum will twist and turn multiple times, while the longer arms make a single, solemn sweep. Variability, in other words, a stunning amount of it, comes built into the El Niño cycle.

Among other things, El Niños and La Niñas can begin in April or June or September, though earlier in the year is more common. They can last for as few as five months or keep going for two or three years. Sometimes several El Niños might occur in a row; then that side of the cycle might take a long break; likewise for La Niñas. Moreover, while there is a preferred order to the swings (the tropical ocean does not usually tarry long at the point called normal), there is no requirement that every El Niño be followed by La Niña, or vice versa. "Summer," in other words, may come once after spring and again after fall before winter ever makes an appearance.

Yet plotted over a long enough period of time, the back-and-forth swings of the El Niño cycle stand out as a series of spikes that produce a recognizable pattern, rather like a jazzman's improvisations on a musical theme. "Quasi-periodic" is the name scientists apply to such behavior, and in this case quasi-periodic means that the usual repeat interval is between two and seven years.

Just as the seasonal cycle imprints itself upon diurnal swings from day to night, so the El Niño cycle imposes itself on seasonal patterns of temperature and precipitation. It does not change winter into summer or vice versa; rather, it alters the character of those seasons by rearranging the large-scale structure of the atmosphere. Thus, El Niño steers torrential rains toward the desert coast of Peru and away from the southern highlands; La Niña flips things back. El Niño shifts rainfall away from normally wet Indonesia; La Niña targets the already arid American Southwest. El Niño protects Florida from Atlantic Ocean hurricanes; La Niña does precisely the opposite.

All this seems so obvious now, as do the profound effects of the El Niño cycle on both marine and terrestrial ecosystems. But for a very long while, the outlines of this powerful climatological upheaval lay buried in a quagmire of regional and local detail. Even today it seems almost preposterous that the expansions and contractions of a pool of warm water could be responsible for so much: not just floods and droughts, landslides and fires, but also the collapse of commercial fisheries and outbreaks of virulent disease.

Before scientists could imagine the El Niño cycle, they first had to imagine the existence of something large enough to sustain such a cycle, and that something turned out to be the Southern Oscillation.

THE SOUTHERN OSCILLATION _____

What is El Niño? Strange though it seems, the key that unlocked the puzzle came from colonial India. In 1875, in response to the dreadful famines that had accompanied a succession of droughts, the organization that later became the India Meteorological Department was set up in Calcutta, the crowded, colorful city that, for a time, served as the capital of the British raj. Then came the drought and famine of 1876–78. Among those stunned by the millions who died was Henry Francis Blanford, the first director. As the casualties mounted, Blanford became more and more determined to ferret out the meteorological cause.

Early on it had become clear to Blanford that India, during this critical time, had come under the sway of abnormally high barometric pressures, and these appeared to him to be the key to the puzzle. By 1880 he had figured out that the trouble in India was part of a much larger disturbance that had rolled over the entire Indo-Australian region.

Over this great expanse, noted Blanford in an official report, barometric readings reached their "maximum between July and September 1877, due to the condition of the higher strata of the atmosphere," and were "most intense, as far as is presently known, in South Australia; and in the Indian region, on an axis extending from the Nicobars to Bengal." The disturbance was "perceptible as far east as Zi-ka-wei in Northern China and Sydney in Australia," but it was "less intense at Ceylon than at the Nicobars and in the Western Himalaya than in Sikkim."

The reports coming out of China in 1877–78 were, in fact, every bit as horrifying as those from India. Wrote the chairman of the Foreign Relief Committee: "The autumn crops over the whole of Shansi and the greater part of Chihli, Honan and Shensi had failed. During the winter and spring of 1877–78, the most frightful disorder reigned." The road to Shansi, he added, "was covered by the carcasses or skeletons of men and beasts, and the wolves, dogs and foxes soon put an end to the sufferings of any wretch who lay down to recover from or die of his sickness."

Shortly after leaving the city of Shanghai, another eyewitness, Fredk. H. Balfour, reported,

The people's faces are black with hunger; they are dying by thousands upon thousands. Women and girls and boys are openly offered for sale to any chance wayfarer. When I left the country a respectable married woman could be easily bought for six dollars and a little girl for two. In cases, however, where it was found impossible to dispose of their children, parents have been known to kill them sooner than witness their prolonged sufferings, in many cases throwing themselves afterwards down wells, or committing suicide by arsenic.

Scientists, of course, had known about the connection between atmospheric pressure and rainfall for more than three centuries, practically since 1604, when the Italian scientist Evangelista Torricelli invented the mercury-filled barometer. "What a marvel it is," Torricelli wrote, "that the quicksilver rises to the point at which it is in balance with the weight of the . . . air that is pushing on it." It was only during the latter part of the nineteenth century, however, that they began to gather inklings that pressure fluctuations in one large region of the world might be connected to those far distant.

To Blanford, the thought of asking Australia and other regions for barometric readings seemed as natural as breathing, for like others of his era, he had been steeped in the global habit of mind fostered by the British Empire. It was not, however, just the political or geographic scope of the empire that invited such expansiveness of thought. After all, the Romans had enjoyed that much. What was far more important was the series of extraordinary inventions that underlay the British Empire and brought its disparate pieces so artfully together.

For the widespread perception—indeed, one might say, the illusion—that India and Australia, though separated by thousands of miles, are, in other ways, not so far apart did not begin in the twentieth century but in the nineteenth, and it began with the invention of the steam engine and the electric telegraph.

Even in the time of Thomas Jefferson, historian Stephen Ambrose has delighted in pointing out, information was forced to travel at the speed of a sailing ship or a man on horseback. In such a world, the distance between India and Australia could be more accurately measured in time than in distance. Even the most urgent questions either received no answers for months or, worse yet, answers meant for earlier questions.

But by the end of the nineteenth century, all that had changed. The invention of the steam engine led to the construction of railroads and the establishment of shipping routes that sped information across continents and oceans. The invention of the electric telegraph made it possible to transmit information from the far corners of the globe in almost no time at all.

It was over the telegraph that Blanford had broadcast his request for information about weather patterns in the other countries that bordered the Indian Ocean. Among those replying was Charles Todd, Australia's postmaster general and architect of the Overland Telegraph. Between 1870 and 1872, Todd and his men managed to string a single iron wire across 2,000 miles of some of the most forbidding country imaginable, including the harsh red-rock desert around Alice Springs, which bears the name of Todd's wife.

The overland portion of the line ended at Darwin, a rough, rawboned port bordering the Timor Sea. From Darwin, the telegraph continued underwater to Java, and when the first message sped through in 1872, Todd rejoiced: "The Australian colonies were connected with the grand electric chain which unites all nations of the earth."

In 1879, Todd enthusiastically replied to Blanford's request and supplied him with all sorts of timely weather information, not only from southern Australia but also from the once remote Northern Territories, including Darwin, whose post and telegraph station commanded a sweeping view of the horseshoe-shaped harbor in Beagle Bay, named after the fabled ship that carried Charles Darwin around the world.

A program of meteorological observations began at the station around 1872, before it moved from a ramshackle wooden structure to a more solid complex of mortar and stone, and continued until World War II, when weather operations moved out to a more strategic location on the edge of the military airfield. The post and telegraph station was, in fact, destroyed by Japanese bombers, but amid the rubble the station's old-fashioned mercury barometer stubbornly clung to its place on the wall.

It was only much later that the importance of the pressure record at Darwin would be appreciated; in 1879 it was only one of perhaps a dozen such records that Todd relayed to Blanford, including records from Mauritius and New Zealand. Anomalously high pressure, Todd reported, was hardly unique to India, but prevailed over the entire region, with the result that rainfall was greatly reduced and many areas found themselves in the grip of extended drought. By the time the drought had run its course, ranchers in New South Wales had lost five million sheep.

Todd went on to suggest to Blanford that a monthly exchange of weather information between India and Australia might prove useful "to facilitate investigations into the distributions of pressure and the origin and march of atmospheric disturbances." With those words, Todd could be said to have anticipated the quest that would lead, some decades hence, to Sir Gilbert Walker's discovery of a giant atmospheric seesaw whose fulcrum lay in the tropical Pacific somewhere between Darwin and the island of Tahiti.

GILBERT THOMAS WALKER was born, on June 14, 1868, the first son and fourth child in a family of seven children. According to a biographical memoir published by the Royal Meteorological Society, his father, Thomas Walker, was the borough engineer in Croydon and apparently quite expert in building reservoirs and the like out of that sturdy material reinforced concrete. His mother, the memoir goes on to note, became a "permanent invalid when Gilbert was about six years old," but no reason for her illness was given. Perhaps he was tutored at home initially, since it was not until 1876 that he entered the Whitgift Grammar School in Croydon. It is tempting to speculate that perhaps he might have been one of thousands of schoolchildren across England who are said to have dipped into their pockets to contribute to the India Famine Relief Fund in 1877.

From Croydon, Walker went on to St. Paul's, where he early on gravitated to the study of mathematics. He also exhibited considerable talent for construction and design. At age seventeen, he fashioned a gyroscope that won him a prize. A short time later, he entered Cambridge University on a mathematics scholarship. It was as a Cambridge undergraduate, apparently, that Walker became fascinated by boomerangs and so expert at throwing them that he became known to his friends as "Boomerang Walker."

He excelled at his studies, winning high academic honors, though according to the Royal Meteorological Society, "the hard work necessary to attain these academic successes so impaired Walker's health that it broke down in 1890 and he had to spend the three subsequent winters in Switzerland. These interludes, however, were valuable to him since, besides giving him time to think, they provided the opportunity for him to become an expert skater, and skating became one of the greatest pleasures in his life."

Walker was a man of versatile talents, and he arranged his professional life so that it embraced every interest, every hobby. He was intrigued by bird flight and hence contributed to the Royal Aeronautical Society an article entitled "A Note on the Wings of Gliding Birds." He played the flute and thus came to write a paper entitled "The Theory of the Flute." Things that moved fascinated him, bicycles and boomerangs and golf balls and spinning tops, and he wrote about those as well.

"Boomerangs," he observed in an engaging treatise on that topic that appeared in the journal *Nature* in 1901, "may be studied for their anthropological interest as examples of primitive art or for the manner in which they illustrate dynamical principles. But there is extraordinary fascination in making and throwing them, and in watching the remarkable and always graceful curves described in their flight."

Walker, in fact, made his own boomerangs, carving them from blocks of ash and taking care that "the outer edge . . . should follow the grain of the wood." Some he no doubt decorated with attractive designs, as Walker, in addition to his other abilities, was said to be an accomplished artist, much drawn to watercoloring.

Between 1895 and 1902, Walker held the position of lecturer in mathematical physics at Cambridge. But in 1903 he resigned that post to go to India as assistant to Sir John Eliot, who had succeeded Blanford as meteorological reporter to the government of India and director general of Indian observatories. Eliot, apparently, had come to England in order to handpick his own successor, though how he came to meet Walker is unclear. "It was a surprising but very wise choice," observed George Simpson, whom Walker hired in 1906. "Surprising because Walker was a typical Cambridge don and had never read a word of meteorology; and wise because he chose a man full of energy, ready to learn. . . ."

And wise also, one might add, because the atmosphere is perhaps the ultimate dynamical system whose multiple moving parts are ruled by the same laws of physics that Walker so loved to contemplate.

FROM 1904 UNTIL 1924, when he retired and returned to England, Walker lived in Simla, a fairy-tale village surrounded by the dramatic spires of the high Himalayas. As it happened, Simla, which Walker made the headquarters of the India Meteorological Department, was also the summer headquarters of the British raj.

There Walker would sometimes impress a celebrity audience with his mastery of the boomerang. In attendance, usually, were the viceroy of India and a coterie of his friends. It was said that Walker could throw a boomerang so that it described not two but three wide circles in the air before returning to the ground at his feet. It was also said that he could throw a boomerang so that it executed a perfect figure eight.

Walker, though, had not come to India to amuse others with his antics but to solve the mystery of the monsoon. For a time, his predecessor, Sir John Eliot, had appeared to make substantial progress in this regard, issuing several forecasts that appeared reasonably on target. But Eliot's forecast system (which included an aptly chosen grab bag of considerations, from the amount of snow cover in the Himalayas to the height of the annual flood along the Nile) broke down in the 1890s. Indeed, Eliot utterly failed to anticipate the extended drought that struck India in 1899, triggering yet another appalling famine.

Eliot worried a great deal over this failure, which he attributed to special meteorological conditions that prevailed between 1892 and 1902. During this period—"the unique period," he termed it—"various parts of the Empire, more especially India, Australia, and South Africa, have suffered more severely from droughts and famine than during any other similar period for a hundred years at least."

Presciently, the idea of a pressure "seesaw" appeared in Eliot's writings, as did the idea of including data from other areas of the world. "The inquiry into the meteorology of this unique period has not proceeded beyond this stage in India for lack of data," he observed. "It would be very desirable that information for Australia, South and East Africa as complete as that for India should be forthcoming."

The suggestion that meteorological patterns were somehow coupled on very large scales appealed to Walker, who came of age during an era when scientists had begun to tease apart the complexity of the biological, physical, and chemical worlds, revealing hidden connections. It was during the latter half of the nineteenth century that James Maxwell demonstrated that electricity and magnetism were but two expressions of the same force, that Charles Darwin revealed how all living creatures could have evolved from a common ancestor, that Dmitri Mendeleyev arrayed the known elements on the periodic table and showed how the metals and the noble gases were, at a deep level, related.

It was also during the latter half of the nineteenth century that new ideas about climate exploded, including the intriguing hypothesis that the eleven-year sunspot cycle was the controller of drought-rain cycles around the world. The latter half of the nineteenth century also saw validation of the controversial idea put forward most provocatively by Swiss geoscientist Louis Agassiz that much of the earth had once been locked in a vault of ice.

In 1909, Walker began to publish a stream of papers on Indian weather, among them papers that considered possible correlations between the strength of the monsoon and variations in solar luminosity. But while Walker never came up with evidence of a strong connection between the monsoon and sunspots, he did identify a surprisingly strong statistical relationship between rainfall in India and atmospheric pressure readings across a breathtaking geographic expanse, from Santiago, Chile, at the eastern edge of the South Pacific to the island of Mauritius in the Indian Ocean.

Scientifically, some of Walker's most productive years came during World War I, when the other senior scientists in his office were assigned to military duties. "So, to find full occupation for members of junior staff," one of his colleagues recalled in a 1953 tribute, "Walker arranged this long-term programme of computation." What this entailed was not computing in any modern sense but rather combing through pages of meteorological data collected from around the world and, then, by dint of statistical sleight of hand, uncovering heretofore unsuspected relationships.

In later years, Walker was often held up as an example of a scientist who had become so enamored of statistics that he found correlations everywhere, even when none existed. But Walker himself was acutely aware of such pitfalls and took pains to avoid them. "In making meteorological applications of statistical methods," he cautioned, "it is important to have as definite ideas as possible regarding the reliability of the conclusions reached, and of the causal relationships which the analysis may indicate."

The crowning achievement of Walker's career came in two landmark papers, one published by the India Meteorological Department in 1923, the other in 1924, the year he returned to England. He drew upon data from India, Australia, Africa, and the Americas in a search for the grand undulations in pressure—"swayings" he called them—that he believed controlled the vagaries of weather worldwide.

In the first paper, Walker remarked on not one but three oscillations in atmospheric pressure that he considered important. "We can perhaps best sum up the situation by saying that there is a swaying of press [ure] on a big scale backwards and forwards between the Pacific Ocean and the Indian Ocean, and there are swayings, on a much smaller scale, between the Azores and Iceland, and between the areas of high and low press [ure] in the N. Pacific."

In the paper published a year later, Walker gave these swayings names: the North Atlantic Oscillation, the North Pacific Oscillation, and most important of all, the Southern Oscillation. "We can perhaps best sum up the situation by saying that by the Southern Oscillation," he wrote, "is implied the tendency of pressure at stations in the Pacific . . . to increase, while pressure in the region of the Indian Ocean . . . decreases."

AIR, OF COURSE, flows from high pressure to low, so that what the Southern Oscillation described was the slip-sliding of an ethereal continent composed of quadrillions of atoms and molecules. In addition to oxygen and nitrogen, air contains methane and carbon dioxide, the greenhouse gases that trap heat from the sun and prevent it from radiating out into space; air contains water vapor that condenses into cloud droplets that refract sunlight, sometimes producing a rainbow of colors; air contains dust particles of infinitesimal size. If air appears insubstantial, it's only because it's so diffuse; compressed, it would turn out to be surprisingly weighty stuff.

Assisted by a younger colleague, Walker spent much of the rest of his life refining the ideas presented in the 1923 and 1924 papers and amassing more evidence to support his thesis. He had a practical bent of mind and liked to use the term *foreshadowing* when envisioning the use to which his research might eventually be put.

"In general the object of prediction," he observed at a 1930 meeting of the Royal Meteorological Society, "is to assist the layman, and it is the opinion formed by him that decides whether they will succeed or fail. Hence I regard it as foolish to issue a prediction except in years when the indications for an excess or deficit [of rainfall] are so strongly marked as to give a 4:1 chance of success . . . and as the claim to 'forecast' the seasons arouses the expectation of an annual precipitation, I advocate the word 'foreshadow' as expressing a smaller ambition."

Today Walker's ideas spin out across time like the boomerangs he so loved to throw. But in his own day, the pioneering work he had done seemed only to diminish in importance. As an obituary published by the *Quarterly Journal of the Royal Meteorological Society* in 1959 poignantly noted, Walker was driven to undertake "a statistical examination of the relation between events, in many parts of the world, which might or might not be expected to show any physical connection. . . . [His] hope was presumably not only to unearth relations useful for forecasting but to provide a productive starting point for a theory of world weather. It hardly seems to be working out like that."

Out of kindness, perhaps, and because he also admired Walker, the writer of the obituary went on to caution against too hasty a judgment, adding that as the science of meteorology progresses, "it may well be profitable to look back and see whether Walker's findings give a lead for further advances."

Walker published his last scientific papers between 1939 and 1947; the topics covered included the effect of Arctic cold on the world's weather and the impact of clouds on the popular sport of gliding. (In old age, his main regret, apparently, was that slow reflexes due to advancing age prevented him from becoming a glider pilot.) Some years later, a former colleague recalled running into him; Walker, by then well into his eighties, was on his way to a flute lesson and caught up in writing a book about the history of the instrument and its design.

Sir Gilbert Walker died on November 4, 1958, at the age of ninety. By then, other scientists, working on what appeared to be an entirely separate problem, had taken the first steps along the trail that would lead them to connect the odd ocean warming off South America known as El Niño to the planetwide circulation feature known as the Southern Oscillation. Among these scientists was Jacob Bjerknes.

THE SPUR TO SYNTHESIS _____

In the fall of 1957, the coral ring of Canton Island, in the memory of man ever bleak and dry, was lush with the seedlings of countless tropical trees and vines. Two remarkable and unprecedented events gave rise to this transformation, for during 1957 great rafts of sea-borne seeds and heavy rains had visited her barren shores. One is inclined to select the events of this isolated atoll as epitomizing the year, for even here, on the remote edges of the Pacific, vast concerted shifts in the oceans and atmosphere had wrought dramatic change.

Elsewhere about the Pacific it also was common knowledge that the year had been one of extraordinary . . . events. Hawaii had its first record typhoon; the seabird-killing El Niño visited the Peruvian Coast; the ice went out of Point Barrow at the earliest time in history; and on the Pacific's Western rim, the tropical rainy season lingered six weeks beyond its appointed term.

From the January 1960 *California Cooperative Oceanic Fisheries Investigations Reports*

THE HISTORY OF science brims with serendipities, and the way Jacob Bjerknes latched onto the problem of El Niño surely qualifies as one of these, as happy and extraordinary as a chance encounter that produces a lifelong romance.

"In the late 1950s," remembers University of Washington oceanographer Warren Wooster, now semiretired, "Bjerknes was at UCLA, as was his colleague Morris Neiburger. Neiburger had a contract with the Inter-American Tropical Tuna Commission, whose director, Milner Schaefer, believed that the odd ocean warming known as El Niño had some effect on the abundance and distribution of tropical tuna. As it turned out, Neiburger had other commitments that prevented him from working on the problem, so he turned the matter over to Bjerknes."

From that moment, believes Wooster, the insights that were to pour out in the years to come became almost inevitable, for the match between the puzzle posed by El Niño and a mind accustomed to thinking on global scales proved to be synergistic beyond anyone's imagining.

Jacob Bjerknes, "Jack" (pronounced Yock), as his friends at UCLA called him, represented the third generation of a scientific dynasty. He was the grandson of mathematician Carl Anton Bjerknes and the son of physicist Vilhelm Bjerknes, who, in 1918, founded a geophysical institute in Bergen, Norway, which quickly became the world's leading center for cutting-edge meteorological theory.

As a young man, Jacob Bjerknes generated many of the ideas that undergirded what came to be called the frontal theory of weather. Structurally speaking, tropical storms were simple beasts. To explain a hurricane, for example, one needed little more than a source of heat and moisture plus the spin provided by the earth's rotation. Extratropical cyclones, the class of storms that brought so much heartache to California in 1998, were more complicated; they demanded additional explanation.

What Bjerknes added was the understanding that these storms arise from the clash of cold and warm fronts. It was this insight, in fact, that led to modern weather forecasting and to the familiar weather forecast maps that now parade across nightly TV newscasts in much of the world. Indeed, by the 1920s, when Sir Gilbert Walker was starting to publish his most important work, the ideas spilling out of Bergen, Norway, were generating such an enormous swell of excitement that they eclipsed practically everything else. For many years thereafter, they dominated meteorology.

It was thus a delicious irony that Jack Bjerknes played such a key role in the rediscovery of Sir Gilbert Walker's pioneering work, that he recognized the worth of it the way a savvy art dealer might if he were to come across a dusty Rembrandt in a pawnshop. For while other scientists of his generation—Hendrik

Petrus Berlage of the Netherlands, Alexander James Troup of Australia—also understood the importance of the Southern Oscillation, Bjerknes was the first to explain its strong physical link to El Niño.

Initially, what sent Bjerknes rummaging around in odd, forgotten corners was a desire to understand the ocean's role in regulating climate. Quite naturally, he began with the North Atlantic. But, around 1957, he expanded his search to include the tropical oceans as well. Not coincidentally, the period between July 1957 and December 1958 was officially designated as the so-called International Geophysical Year, and the coordinated observations it occasioned documented such turmoil around the world that meteorologists and oceanographers alike were forced to think in new directions.

A symposium on "The Changing Pacific Ocean in 1957 and 1958," held in Rio Rancho, California, in June 1958, captured this extraordinary moment. Bjerknes himself does not appear to have attended, but his former student Jule Charney did, and so did other leading meteorologists, oceanographers, and fisheries biologists, including Warren Wooster, then at the Scripps Institution of Oceanography in nearby La Jolla, and the Inter-American Tropical Tuna Commission's Milner Schaefer.

Even today, the notes from that meeting convey the excitement as well as the confusion of scientists tentatively approaching something very large and very strange. Charney, one of the early pioneers of computer-based weather forecasting, chaired the session. "I wish to say here that we meteorologists have been particularly at fault in regarding oceans as a kind of passive body which merely responds to motions in the atmosphere," he observed. "It is clear, if one considers long-period changes, that this cannot be the case. Indeed, for long-term changes, it is probably more the other way around. Ultimately we shall have to consider the oceans and the atmosphere as a coupled dynamical system."

What astonished the scientists meeting at Rio Rancho was the widespread nature of the ocean warming that occurred in 1957–58. Because of the International Geophysical Year, oceanographers had arranged for commercial ships to take monthly temperature readings. These readings unequivocally established that the entire eastern basin of the equatorial Pacific had warmed, from the coast of Peru to little Canton Island, which lies just east of the International Date Line.

Warming also occurred off the California coast; during the summer of 1957, in fact, sea-surface temperatures had soared to their highest levels in twenty-one years, so that "more than four times the solar heat actually received would have been necessary to account for the warming," one symposium participant marveled. Species of tropical fish, including skipjack tuna and dolphin fish or mahimahi, followed the warm water. In 1957, California recreational fishermen hooked an astonishing 3,000 dolphin fish, compared to the previous record of only fifteen landed a decade earlier. If this sounds familiar, it should. Virtually the same thing happened during the El Niño of 1997–98.

And then there was Peru, where once again anchoveta became scarce and seabirds starved en masse. In 1956, in fact, Peru boasted a total population of some 27 million pelicans, cormorants, and boobies; in 1958 the population stood at 6 million. For the first time, the biological upheaval in the ocean so vividly chronicled by ornithologist Robert Cushman Murphy appeared connected, in some way, to a much larger climatological upheaval.

By the time the Rio Rancho symposium adjourned, it seemed clear to the participants that they were on the verge of gaining important new insights into the earth's climate system. Warren Wooster, in particular, was intrigued by the similarities between the upheaval in Peru and the upheaval off the California coast and even went on to speculate that the underlying physics might be similar. But neither he nor anyone else understood the degree to which the disturbances around the world in 1957 and 1958 might be linked until Bjerknes began zooming in on El Niño with his high-powered lens.

BJERKNES PUBLISHED HIS first paper on El Niño in 1961. "The meteorological control over the occurrence of El Niño must lie in the fluctuating strength of the Pacific tradewinds," he observed. "Ocean-wide weakness of the . . . tradewinds would permit a larger-than-normal volume of warm water to assemble in the Eastern Tropical Pacific." Bjerknes, in other words, redefined the problem of El Niño as a very large scale response on the part of the ocean to the waxing and waning of one of the world's great wind systems.

Bjerknes then zeroed in on a puzzle that had intrigued him for several years: why tiny Canton Island, usually so dry, turned so extraordinarily wet in 1957. Named for an American whaling ship that foundered off its coast, Canton (now respelled Kanton) is little more than a narrow rim of coral-encrusted rock near the center of the equatorial Pacific. Not counting the lagoon, it covers an area of 3.5 square miles. During World War II, Canton served as an Allied air base; for a time afterward, commercial airliners used it as a refueling stop on long Pacific flights.

What Bjerknes noticed was this: The years of extreme rainfall recorded on Canton seemed to coincide with notable El Niños as documented in Peru. "Big monthly totals of rain can be seen to occur only during periods when the ocean is warmer than the atmosphere," he observed in the *Monthly Weather Review*. "Under those conditions, the maximum upward transfer of moisture takes place, and an intense convection brings the moisture to the level of condensation and farther up into towering shower clouds."

In the course of researching the scientific literature, Bjerknes connected with the writings of Sir Gilbert Walker, and in the uncanny way that one great intellect frequently resonates with another, Bjerknes recognized Walker's genius. To Bjerknes, the statistical manipulations that Walker had engaged in were flares that illuminated an extraordinary phenomenon, something that neither meteorology nor oceanography alone could adequately explain.

He began filling in the picture. Walker had imagined a swaying of pressure in the atmosphere; Bjerknes paired it with a thermal commotion in the ocean. This commotion was caused by the waxing and waning of the winds, the "Trade Winds," as Ishmael, the narrator of *Moby Dick,* observed, "that in clear heavens blow straight on, in strong and steadfast, vigorous mildness; and veer not from their mark, however the baser currents of the sea may turn and tack." It was the dying of these usually dependable winds—"bodiless as objects, but not as agents," declared Ishmael—that gave birth to El Niño, Bjerknes realized, and their refreshing resumption that signaled its death.

What mattered to the trade winds was the gradient in atmospheric pressure on opposite sides of the ocean, the very gradient that the Southern Oscillation described. When pressure was high in the eastern equatorial Pacific and low in the west, the gradient was steep. In essence, it formed an invisible slope down which the trade winds swooped like sledders on an icy hill. But as pressure in the east dropped, the gradient flattened out. At that point, the winds slowed down and even reversed direction.

To Bjerknes this seemed significant. The trade winds, scientists by now clearly understood, were the agents that made Peruvian waters so cold and, also, so biologically rich. Channeled by the Andes, the trades swept up from the south, paralleling the coast until they reached the equator and swerved sharply west. All along their land-hugging route, these winds shoved water aside, creating a coherent offshore flow. To fill the long trough that the winds created, cold water, brimming with nitrogen and other growth-promoting nutrients, welled up from the depths below. It was the weakening of the trades that shut down the upwelling of this cold, abyssal water, Bjerknes reasoned, and hence, the weakening of the trades that caused the ocean to warm.

Connecting the ocean to the atmosphere, El Niño to the Southern Oscillation, was a pattern of airflow that Bjerknes named the Walker Circulation. An oft-cited analogy invokes the flow of heated air through an imaginary "room," only in this case the floor happens to be the skin of the ocean, the ceiling, the underbelly of the stratosphere.

Warmed by a basement "furnace" (in this case a vast pool of sun-warmed water), air flows out of a vent, then shoots up and spreads out across the ceiling. But as it rises, the air that started out so warm and buoyant becomes colder and therefore heavier. Rather like a hot-air balloon that has lost its oomph, a parcel of air spreading across the ceiling sags and sinks back to the floor. At that point, it joins a room-size draft (the trade winds) blowing back in the opposite direction. For this very reason, home-heating engineers generally position vents for return flow on or near the floor. In this way, cold air that scoots along at low levels is continuously recaptured and rerouted to the furnace for heating.

The image of air flowing through a room is useful because it provides a construct that encompasses both "convection" and "subsidence." Thus, the rising air associated with convection leaves behind a zone of low pressure near the floor; subsidence, by contrast, is characterized by down-falling air and high pressure near the floor. At this point, though, the analogy breaks down unless one cares to hook the furnace to a humidifier so large that it would be capable of producing rain.

For the heat trapped in warm water powers a tremendous amount of evaporation, and so the air that rises from the equatorial ocean is wet as well as warm. As the air rises and cools, the moisture it contains will condense out as cloud droplets and, ultimately, as rain. The wall of rising air in the imaginary room described by the Walker Circulation, in other words, is also a wall of rain, a thick, continent-size wall of rain.

By studying satellite images of cloud cover, Bjerknes figured out that, in most years, that wall loomed over the Indonesian Archipelago. In El Niño years, however, it shifted to the east, leaving western Pacific islands like Borneo and New Guinea in drought and turning the region from Canton Island to Peru's coastal desert almost unimaginably wet.

Embedded in the concept of the Walker Circulation, indeed integral to it, was a paradox akin to that posed by the chicken and the egg. Bjerknes himself could not specify which came first, the disturbance in the atmosphere or the disturbance in the ocean. Just as the swaying of atmospheric pressure governed the strength of the trade winds and the waxing and waning of the trade winds determined the temperature of the ocean, so, too, the temperature of the ocean determined the swaying of pressure and thus the strength of the winds.

What Bjerknes envisioned was a "chain reaction" in which the atmosphere and the ocean ran off in one direction, then reversed course and ran off in another. "There is," he mused, "ample reason for a never-ending succession of alternating trends by air-sea interaction in the equatorial Pacific, but just how the turnabout in trends takes place is not yet quite clear."

Initially, the ideas Bjerknes outlined received mixed reviews. Meteorologist Eugene Rasmusson of the University of Maryland wryly remembers his own reaction. While he was studying atmospheric science at M.I.T., Sir Gilbert Walker was often cited as a cautionary example, proof of how easy it was to misapply statistics and indulge in spurious correlations. "So when I saw that paper by Bjerknes," Rasmusson said, "I thought to myself, 'That poor old man, he's gotten tangled up with Walker. His reputation will be ruined.' "

Not long afterward, Rasmusson changed his mind about Bjerknes and about Walker as well. In 1983, he and a colleague, John M. Wallace, coauthored a paper on the climate cycle that Walker and Bjerknes and those unnamed sailors from Paita could be said to have codiscovered. The El Niño Southern Oscillation, they and others now called it, or ENSO for short.

Walker glimpsed the great flukes splashing down on the ocean; the Paita sailors first caught sight of its spume. But it was Bjerknes who saw both the head of the beast and its tail, and so it was Bjerknes who assembled the first compelling portrait of El Niño, a chimerical creature spawned by the passionless coupling of wind and water, a force of nature wilder and more elusive even than the White Whale. As a result, the picture of El Niño that scientists carry around in their heads is still largely the one that Bjerknes impressionistically painted, though it has been greatly embellished, and strengthened, by insights scientists have since gleaned.

CHAPTER **THREE**

But not yet have we solved the incantation of this whiteness, and learned why it appeals with such power to the soul. . . . Is it that by its indefiniteness it shadows forth the heartless voids and immensities of the universe? Or is it, that as in essence whiteness is not so much a color as the visible absence of color, and at the same time the concrete of all colors; is it for these reasons that there is such a dumb blankness, full of meaning, in a wide landscape of snows?

HERMAN MELVILLE, *Moby Dick*

QUELCCAYA

More than a quarter century ago, on a sun-drenched morning in June 1974, Lonnie Thompson climbed into a battered old truck and made the first of many journeys from the bustling market town of Sicuani to a little village called Soccopalca. There, in one of the high valleys that cut through the Peruvian Andes, the road ended, and the twenty-six-year-old graduate student from Ohio State University emerged from the truck to find himself standing on the edge of a rock-strewn meadow stippled with clumps of straw-colored grass.

Night was only about an hour away, and already the day's warmth was beginning to fade. Some distance from a small cluster of simple stone houses, Lonnie and his three companions set up their tents, ate a simple dinner, and tunneled into their sleeping bags. The next morning, they were hoping to get off to an early start.

They planned to be out in the field for a total of about six weeks and so had brought with them a fairly hefty load of supplies. They had packs crammed with camping equipment; duffel bags filled with ropes, ice axes, and crampons; shovels for cutting into ice. They had burlap sacks stuffed with potatoes, onions, cans of meat and fruit, packages of cereal, pots, pans, and other sundries.

It was far too much for them to carry in on their backs, and so they had made arrangements for packhorses to meet them. But the packhorses did not arrive at Soccopalca the next morning, nor did they arrive the morning after that, and to pass the time, Lonnie and the others began to explore the surrounding countryside. Separately and together, they wandered along the splashing waters of the Salcca River, fanned out across sloping grasslands browsed by mixed herds of alpaca and sheep, and climbed some of the rocky prominences that stood sentinel above the valley.

It was on top of one prominence, as he drew his breath in ragged gasps, that Lonnie caught his first glimpse of the Quelccaya Ice Cap. Broad and very nearly flat, but with steeply angled sides, the ice cap resembled a dazzling white coverlet thrown over a slightly rumpled bed. Only in this case the coverlet measured 22 square miles and spread across an 18,700-foot-high plateau framed by jagged snow-clad peaks. Even at a distance, it seemed enormous.

At that moment, Lonnie did not realize it, but he would spend the next decade of his life wrestling with that ice cap—with Quelccaya's remoteness, its ruggedness, and above all, its height. It would be an epic match, one that Denmark's Willi Dansgaard, one of the titans in the exploding field of paleoclimatology, predicted Lonnie would lose. Quelccaya was too high for humans, Dansgaard said.

But all that lay ahead, as did the El Niño that would change the intellectual landscape of climate science as profoundly as the rise of the Andean cordillera had altered the topography of Peru. As Lonnie would come to appreciate, these two upheavals were not as unrelated as they might seem. It was the rise of the cordillera that helped establish the weather patterns so profoundly disrupted by El Niño and ensured that a record of those disruptions would be preserved in Quelccaya's ice.

In 1974, however, Lonnie had not yet made the connection between Quelccaya and El Niño. In fact, prior to making that fateful trip to Peru, he had not even heard of El Niño, and certainly he had no abiding interest in the tropics. He was a glaciologist, and along with virtually everyone else in the field, he shared an unexamined belief that the torrid midsection of the world was, climatologically speaking, inert. The action, it was all but universally assumed, lay in the polar regions, in the dramatic expansions and contractions of the great ice sheets whose impressive remnants continued to envelop the earth's highest latitudes in razor-edged cold.

But Lonnie was young, and his preconceptions had not yet hardened into prejudice. So when an older scientist at Ohio State's Institute of Polar Studies started spinning tales of an ethereally high ice cap some 800 miles south of the equator, Lonnie expressed interest, even enthusiasm. The older scientist's name was John Mercer. He was an expert on glacial moraines and something of a free spirit as well. While most of his peers remained content to work on the well-chronicled climate history of the Northern Hemisphere, Mercer had branched out. He had tackled the daunting task of documenting the major glaciations of the Southern Hemisphere, about which almost nothing was then known.

On one of his many trips to South America, Mercer had bumped into a Canadian mountaineer named John Ricker, who had newly returned from trekking through the Cordillera Vilcanota, the Andean chain that lies just north of the vast, windswept plateau known as the Altiplano. Not far from 20,945-foot-tall Ausangate, Peru's fourth highest peak, Ricker had spotted a shimmering expanse of snow-covered ice, and even though he had seen the ice cap only from a distance, his report immediately piqued Mercer's interest. Where there was an ice cap, he knew, there were also bound to be moraines that marked a lengthy sequence of ice advances and retreats.

At the time, Lonnie was working on the climate records encoded by ice from Antarctica and Greenland; indeed, for his Ph.D. thesis, he was planning a comparative study of these nearly antipodal regions. It might help fill in the picture, Mercer suggested to his younger colleague, if he also included a sample of ice that was strategically positioned almost exactly in between. As Mercer saw it, Quelccaya's ice might shed light on a central mystery: whether the big advances of ice in the Northern and Southern Hemispheres were synchronous or not.

In the early 1970s, the traditional method of tracking past epochs of glacial advance—Mercer's method—was falling out of fashion, and research grants were starting to reflect that shift. Instead of examining the piles of rocky debris left as the great ice sheets retreated, scientists were increasingly turning to what remained of the ice sheets themselves, and from the crystalline citadels of Antarctica and Greenland, they were beginning to extract a wealth of information. No one, though, had yet considered doing the same thing with tropical ice.

The oversight was the result of benign neglect. Scientists who lived and worked at higher latitudes presumed that ice fields in the tropics underwent such regular melting that their value as climatological archives ranged from little to none—and in most cases this presumption was right. Water trickling through ice is not unlike paint thinner running down a graffiti-covered wall; it diminishes and ultimately destroys the clarity of the calligraphy. Mercer, though, had a hunch that Quelccaya might prove to be different. It was large and flat, just like a polar ice cap, he told Lonnie, and its extreme elevation, 18,700 feet, suggested that it might well remain consistently cold.

As Lonnie looked on, Mercer pulled out a series of aerial photographs that revealed Quelccaya's clean, unbroken surface and contour maps that pinpointed its exact location. In this way, long before he saw it, Quelccaya had begun to loom in Lonnie's imagination.

Lonnie and John Mercer discussed mounting an expedition. Gradually, a plan took shape. To explore the ice cap, Lonnie would team up with Cedomir Marangunic, a Chilean geologist who had studied at Ohio State. Meanwhile, John Mercer, along with John Ricker, would map the moraines in the ice cap's vicinity. They submitted a proposal to the National Science Foundation, and—to their surprise—they received funding, $7,000 to be exact. Initially, the manager of the glaciology program at the National Science Foundation had turned down their request, explaining that the tropics did not fall within his jurisdiction, but at the end of the year, he serendipitously discovered that he had a few thousand dollars more than he thought.

So there they all were in Soccopalca, waiting for the packhorses to arrive. Increasingly impatient with the delay, they began asking the villagers who in the vicinity might have animals for hire. A day or so later, two men, wrapped in dark woolen ponchos and peaked caps called *cholos,* showed up with perhaps eight horses between them. Eager to be on their way, Lonnie and the others hurriedly broke camp, hoisted their daypacks to their shoulders, and left Soccopalca behind.

FOLLOWING THE MEANDERS of the Salcca River, the little procession of men and horses began wending their way up, into the land of perennial frost that begins around 14,000 or so feet. Just beyond their camp, in fact, they crossed over the line drawn by nighttime temperatures that regularly fall below freezing. Beneath this line lay fields of hardy amarga potatoes neatly edged by cobbled walls; above it, only pastureland, cropped close by multicolored alpaca with long eyelashes and quizzical faces and llamas whose ears were bedecked with loops of gaily colored yarn.

Sometimes Lonnie and his companions crossed over beds of volcanic ash or deposits of loess, windblown silt ground by glaciers into fine powder. Sometimes they walked through marshes, using springy hummocks of alpaca moss as stepping-stones. Very occasionally they passed an isolated house with a stone corral and a pyramid of dried peat that, along with llama dung, provided fuel for cooking in these treeless parts.

They were now in the high puna, a minimalist landscape of golden grasses and brooding outcroppings of black basaltic rock. Across it scalding rivers of molten rock and fiery ash had once surged, covering much of the area where Lonnie and the others were now walking. The Quelccaya Ice Cap itself, they would find, lay atop a bed of welded tuff, a dense composite material made out of ash mixed with fragments of basalt and other glassy volcanic rock ejected by a pool of magma buried miles below.

Primordial forces had shaped this landscape and, indeed, were continuing to shape it. At an average rate of more than half an inch per year, the mountains of the Andean cordillera were still rising, shoved skyward by the epic clash between the South American continent and a rogue piece of oceanic crust known as the Nazca plate. The main battlefront lay off the coast of Peru, some 400 miles from Quelccaya, along the spectacular submarine chasm known as the Peru-Chile Trench. Approximately 3,500 miles long, 40 miles wide, and nearly 5 miles deep, this underwater canyon marks a geophysical war zone where volcanoes erupt, earthquakes unleash enormous tremors, and mountains soar to the height of clouds.

As Lonnie soon realized, the place into which John Mercer had drawn him was extraordinary in almost every way, geologically, climatologically, and culturally. Just to the west of Quelccaya lay the coastal desert of Peru; to the east, the emerald fastness of the Amazonian rain forest. To the north lay Cuzco, the former seat of the Incan Empire, and to the south, the deep sky-blue waters of Lake Titicaca, the second largest lake in South America (after Venezuela's Lake Maracaibo) and the highest navigable lake in the world. On the shores of this lake, up until about the year 1100, another remarkable city had flourished. Its name was Tiwanaku, and it was the ceremonial center of another great empire.

The region around Quelccaya was—still is—a crossroads, a junction point, a meeting place. Here warm, moist air from the Amazon meets the cold, dry air of the highlands. Here craggy mountain chains converge in a rocky confluence called the Vilcanota knot. And here Quechua-speaking Indians, descendants of the Inca, mingle with the Aymara-speaking heirs of Tiwanaku.

In this landscape that lies betwixt and between, place names sometimes reflect this dual cultural influence. Titicaca, for example, derives from the Aymara word *titi*, meaning mountain lion, and the Quechua word *caca*, meaning rock. Quelccaya, on the other hand, appears rooted in a word that in both languages means to write or, more broadly, to inscribe. As Lonnie would discover, Quelccaya's dumb blankness had been elaborately written upon by flakes of snow and flecks of dust.

It was late on the first day of the 20-mile hike in, as he rounded the top of a 15,000-foot pass, that Lonnie caught sight of Quelccaya for a second time. To his dismay, it looked no closer than before; like a mirage, the ice cap seemed to have retreated as he approached, matching each of his steps forward with an equal and opposite step back. *When will we ever get there?* he later remembered thinking to himself. He wasn't one to complain, but by this point, he was feeling light-headed, dizzy, and as close to exhaustion as he could ever remember.

The abrupt shift from day to night had already begun. Temperatures that stood at a pleasant 60 degrees Fahrenheit at noon were about to plunge into the twenties. The next morning, Lonnie crawled out of his tent to find hoarfrost on the ground.

Quelccaya did not prove to be as far away as Lonnie had feared. He and the others arrived at its base that same afternoon and pitched their tents between two moraines, near a place where the ice field sloped down in a gradual ramp. They then constructed an open-air kitchen out of a ring of boulders, filling their cooking pots with cold, clear water from a little lake fed at the ice cap's edge. To the northwest, they could make out the stern vis-

age of Ausangate, its precipitous peak enveloped by a black veil
of clouds. That night, and many nights thereafter, the sky around
Ausangate flickered with strobelike lightning.

The next day, Lonnie and Cedomir Marangunic began a series
of methodical explorations, beginning with the perimeter of the
ice cap. A short distance from camp, they encountered an ice wall
that was maybe 60 feet high and looked like an enormous slice of
wedding cake. The ice in the wall was very clearly varved, that is,
layered, and each layer, Lonnie suspected, represented a year's
worth of snowfall. He snapped the series of photographs that
ended up convincing the National Science Foundation that further expeditions to Quelccaya would be well worth funding. "It
didn't take a genius," his wife, Ellen, later commented, "to realize what that stratigraphy meant."

Soon Lonnie and Cedomir ventured onto the ice cap itself,
which, despite its awesome aspect, presented few technical difficulties once they passed the fractured zone at the base. For while
crevasses filigreed the glaciers flowing out from Quelccaya's steep
edges, the interior of the ice cap harbored few such hazards. In
fact, its incline gentled the higher one went, so that it was possible for the two men to trudge the several miles to the summit—or
rather each of three summits—without roping up to guard against
slipping.

The summits were dome shaped, as they were on the Greenland and Antarctic ice sheets, which was an encouraging sign.
Domes mark the highest, coldest spots on an ice sheet. There pressure and temperature are sufficiently balanced so as to allow ice
to build up vertically, as opposed to splaying out in a horizontal
fashion. For ice, even though frozen, retains properties reminiscent of a fluid. Imperceptibly it continues to flow, and it flows
fastest where temperature and pressure are highest.

In an ice cap, as in the ocean, pressure increases with depth, and
for essentially the same reason. In one case, pressure increases because of the weight of the water above; in the other, it increases
because of the weight of the overlying ice. At a certain pressure
point, ice will melt even at temperatures that are nominally below

freezing, and long before that happens, it will turn pliable, like toothpaste, and start to extrude.

To Lonnie, the presence of the domes meant that even at great depths Quelccaya's ice might preserve the annual layering needed to reconstruct a detailed climate record. From that moment on, he stopped thinking of Quelccaya as a tropical ice cap and started thinking of it as a polar creature incongruously transported to a near-equatorial latitude.

Lonnie and Cedomir Maranguníc continued with their reconnaissance. They surveyed the ice cap. They dug a snow pit into the main summit dome and sliced out a sequence of samples from its sides. They shoveled out a sample of snow from 10 feet deeper still. They measured the temperature of the snow, its density, the thickness of the annual layers. An impressive 9 feet, Lonnie estimated, had fallen during the course of the past year. Already it was weighing down on the snow that had fallen in previous years, slowly but inexorably compressing it into ice.

The two scientists hauled the snow samples down from the summit dome in their backpacks. They used the sun's warmth to melt the snow; then they bottled the meltwater. Lonnie loaded the bottles into a duffel bag; he had about four dozen. He was eager to get them back to his laboratory at Ohio State.

It was during the departure from Quelccaya that so many things started to go awry. As Lonnie remembers it, John Ricker decided to go off to make a complete circuit of the ice cap, and when the packhorses came, he had not yet returned. John Mercer declared that the rest of the group would not wait, and they ended up leaving Ricker behind. Then, when they reached the city of Cuzco, Mercer approached Lonnie to see if he could borrow some money. Lonnie was taken aback; he replied that he was very nearly broke. At that point, Mercer said something along the lines of "Well, I'm all out of money, so I guess you'll have to make it back the best way you can."

Lonnie was speechless. He had an air ticket from Cuzco to Lima and from Lima to Miami and from Miami to Columbus, but it was an extended trip that took several days. Where would he stay? What would he eat? He arrived in Lima and threw himself on the mercy of Señor See, a Czech refugee who ran a small pension. He and his wife put Lonnie up for the night and even packed him a box lunch so he'd have something to eat during his long layover in Miami. It was an act of generosity that Lonnie went on to repay a thousand times.

Thanks to the Sees, Lonnie made it to Columbus, Ohio, in good shape, but when he arrived, he was seething. "I wasn't as mad as John Ricker," he humorously comments today, "but I was mad." As he later found out, it wasn't John Mercer's fault that he ran out of money. The older scientist had been robbed very early in the journey, on the train ride in. It was something of a miracle, in fact, that the expedition had continued at all. As soon as Lonnie understood this, his anger subsided. He came to realize, he says, that John Mercer was not unkind, just unlucky. Things like that were always happening to him. "If there were four rafts trying to cross the English Channel," he observed, "John Mercer would be on the one that sank. That's just the way he was."

Lonnie, on the other hand, had an uncanny knack of boarding boats that ultimately made it across, though not without white-knuckle moments. For him, disasters loomed, then dissolved into opportunities. One door slammed in his face just as another swung wide open. That, in fact, is what happened shortly after Lonnie returned from Peru. In this case, it was the door to Greenland that shut, the door to the tropics that opened.

ENTER EL NIÑO

How Lonnie Thompson first heard of El Niño even he can't remember. All that seems certain is that the mere fact Lonnie was traveling through Peru in 1974 virtually guaranteed that El Niño would come up in the course of multiple conversations. Just the year before, the strongest El Niño since 1957–58 had torn through Peru with unprecedented savagery, and the country was still economically reeling. This time around, it was not only marine animals that were adversely affected, but a $350 million industry. By the time that El Niño ended, not just oceanographers, meteorologists, and fisheries biologists but also commodities brokers, agricultural economists, and grain-growing farmers had tuned into the global significance of a phenomenon that, up to then, had seemed little more than a climatological curiosity.

The reason the 1972–73 El Niño had such a disproportionate impact can be attributed to the rise of a commercial anchoveta fishery that had not existed two decades earlier. When the 1957–58 El Niño hit, the teeming shoals of anchoveta off the coast of Peru for the most part still belonged to seabirds, sea mammals, and subsistence fishermen. Ten years later, people by the thousands were leaving their homes in the highlands and pouring into the squalid port of Chimbote, where big steel-hulled ships and stinking seaside plants processed anchoveta by the billions, creating a boom for purveyors of all sorts of goods and services.

The seeds of this economic explosion had been sown very early in the 1950s, when a fish-meal plant surreptitiously sprouted up in a remote, secluded bay. Officially speaking, the plant did not exist; it had been covertly assembled from parts that once belonged to a sardine-processing operation along the California coast. The plant represented a stealth attack by enterprising businessmen whose efforts to set up an anchoveta fishery had been consistently thwarted by the influential families that controlled the harvesting of guano, the dried excrement of cormorants, pelicans, and boobies. In the days before synthetic fertilizer from fossil fuels replaced it in the global agricultural market, guano from Peruvian seabirds, as ornithologist Robert Cushman Murphy put it, was "even more valuable than gold."

What made the guano industry possible was that pelicans, boobies, and cormorants did not defecate in random places, but returned over and over again to the same spots to deposit their strong, ammonia-scented droppings. Over the course of millennia, these birds had erected monumental castles of guano on Peru's offshore islands, particularly the fog-draped Chinchas and Ballestas to the south of Lima. These castles were repositories of biological wealth. "If the value of fertilizer be calculated according to nitrogen content," Murphy wrote, "the best Peruvian guano is three times as effective as farmyard manure."

Guano from the "bird islands" had long been exploited by people. It fertilized the flat, irrigated fields of the Mochica and Chimor Empires and the high terraced fields of the Inca. According to Murphy, the Inca recognized the value of these birds and implemented regulations that protected them, such as putting their islands off limits during nesting season.

In the nineteenth century, commercial interests began strip-mining the vast storehouses of droppings, some well over 100 feet tall and twenty-five hundred years old. Between 1848 and 1875, Peru is said to have exported some 20 million tons of guano worth $2 billion to the United States and Europe for use as agricultural fertilizer. The steep-walled islands the birds called home were sometimes surrounded by dozens of boats, all of them waiting for indentured Chinese laborers to shovel guano into their holds. Virtually no thought was given to the welfare of the Chinese, whose bare backs bore welt marks from whips, and even less thought was given to the welfare of the birds, which were driven off their nests and even directly killed.

Early in the twentieth century, when the deposits of ancient guano had been all but stripped away, the supply of fresh guano suddenly plummeted to a level that threatened the future of the industry. In 1909, at long last, the excesses were curbed and controls established that put a strong emphasis on conservation. In response, the birds staged a robust recovery. In his book *Bird Islands of Peru*, Murphy described the splendid scene:

> Picture to yourself the shining, rainless coast . . . where the swells surge northward, from month to month and year to year. . . . On such an ocean dark flocks of guanayes . . . form rafts which can be spied miles away. Slowly the dense

masses of birds press along the sea, gobbling up fish in their path, the hinder margins of the rafts continually rising in the air. . . . At other times, when the guanayes are moving towards distant feeding grounds, they travel not in broad blocks but as a solid river of birds, which streams in an unbroken column, close above the waves, until an amazed observer is actually wearied as a single formation takes four or five hours to pass a given point.

The birds first sally forth only in small scouting parties, which can be seen flying erratically above the ocean, usually keeping well in the air, and frequently backpedaling or hovering when they see the silvery glint of schooling fish or the ruffled appearance of the sea which indicates the presence of fish below. The dropping of the scouts to the surface, and the shallow dives which mark the beginning of an orgy, are the signals that cause the approach of such rivers of birds as have been described above.

Murphy had long feared that a commercial anchoveta industry would make life difficult for his beloved seabirds, and it appeared he might be right. A breakpoint occurred during the El Niño of 1965–66, when the seabird population, still rebuilding after the El Niño of 1957–58, plunged once more. Five years later, the seabird population was just starting to inch up, but the commercial anchoveta catch was soaring to new heights.

By 1970, two years before the 1972–73 El Niño struck, the Peruvian anchoveta fleet boasted 1,450 steel-hulled vessels that were equipped with echo sounders to locate the anchoveta and strong nylon purse seines to haul them in. The fleet was so large, experts calculated, that in six months it was capable of bringing in 13 million metric tons of anchoveta, nearly twice the quantity that fisheries scientists pegged as the maximum sustainable catch. Processing capacity had more than kept pace. Coastal plants could grind a year's sustainable catch, 7.5 million metric tons of anchoveta, into fish meal in about forty days.

On the eve of the onset of the 1972–73 El Niño, Peru boasted what was easily the most productive fishery in the world. Its catch when the official 1969–70 fishing season ended amounted to 10.5 million metric tons, and even so, many fishermen were dissatisfied. They successfully petitioned the Peruvian government to extend the season, and between that and illicit fishing near the Chilean-Peruvian border, the final catch approached a staggering 11 million metric tons, or 10 trillion individual fish.

And that figure did not include wastage, noted Gerald Paulik, a fisheries expert from the University of Washington. The process of transferring the fish from the hold of the ships (known as *bolicheras*) to the processing plants was notoriously inefficient, he observed. It involved mixing the fish with seawater and sending this piscine slurry through long pipelines that were prone to breaks and leaks. Paulik conservatively estimated the 1969–70 catch to be between 13 and 14 million metric tons, if not higher.

THE 1972–73 EL NIÑO spawned one of the most dramatic reversals of fortune in postconquest Peruvian history. In its wake, the annual catch of the anchoveta fishery plummeted to less than 2 million metric tons and remained low for years. The collapse was as sudden as an earthquake, and like an earthquake, it produced shock waves that sped around the world.

Why? By then, Peruvian anchoveta, dried and ground into fish meal, had become the chief source of high-protein animal feed in international markets, and when the Peruvian fishery foundered, livestock producers in the United States, Europe, and Japan were forced to turn to protein-rich soybeans as substitutes. In May 1973, soybean futures hit a record price of $11.08 a bushel, quadruple what they'd been about a year earlier, and not a few U.S. farmers responded by altering their planting plans, switching relatively more acreage to soybeans versus corn and wheat.

Ordinarily this would not have been a problem for commodities markets to absorb, but it came on the heels of a change in United States farm policy aimed at drawing down a wheat surplus that had become a drag on the prices farmers could fetch for their harvest. As a result, wheat reserves declined sharply between 1970 and 1972, just as serious droughts (many of them linked to El Niño, it would later become clear) affected many crop-growing regions of the world. Among them, the former Soviet Union, Australia, the Indian subcontinent, and Africa's Sahel, the semiarid region that stretches from Senegal to Sudan.

Demand for wheat rose, threatening to outstrip the supply. In the summer of 1973, as international relief agencies warned that as many as 6 million people in the Sahel were in imminent danger of starvation, turmoil seized world grain markets. In the course of a single month, wheat prices in U.S. markets jumped by 80 percent; corn prices, by 30 percent; cattle and hog prices, by more than 25 percent. For a time, as one agricultural economist put it, the entire world seemed to teeter "on a razor's edge with respect to surplus or shortage, feast or famine."

In the end, the collapse of the Peruvian anchoveta fishery did not turn out to be the proverbial missing nail that lost the shoe that lost the horse that lost the rider that lost the kingdom, but it seemed to have come perilously close. To scientists, understanding El Niño suddenly became a lot more important than it had before.

The first attempts to forecast El Niño, in fact, date back to 1974. That was when two oceanographers, William Quinn of Oregon State University and Klaus Wyrtki of the University of Hawaii, independently predicted that an El Niño would occur the following year. Quinn based his prediction on the Southern Oscillation Index, a statistical measure of the atmospheric seesaw in the equatorial Pacific. Wyrtki, by contrast, had devised a more complicated formula that placed greater emphasis on parameters like wind speed and sea level.

That two different prediction schemes had produced such similar results seemed promising, and in the hope they might prove out, the National Science Foundation agreed to fund two El Niño watch cruises. So on February 11, 1975, and again on April 17, a team led by Klaus Wyrtki boarded a University of Hawaii research vessel named the *Moana Wave* and spent the next six weeks observing the eastern basin of the equatorial Pacific first-hand.

At first, it seemed that they were in luck. In February, the ocean east of the Galápagos Islands dramatically warmed. By April, however, temperatures had dropped down again, leading Wyrtki and Quinn to speculate that an El Niño had started but then stopped, suggesting that their approach—while in need of tweaking—might not be so far off track. That perception was reinforced in January of the very next year when a moderate El Niño did commence, so that the predictions made in 1974 seemed to be not so much wide of the mark as tantalizingly close, a pair of bullets that struck the inner ring of a target instead of the central eye.

Looking back, it seems almost uncanny how the El Niño of 1976–77—the El Niño that Wyrtki and Quinn did not quite manage to predict—coincided with a shift of the winds blowing through Lonnie Thompson's professional life. Before it ended, Lonnie would manage to return to Quelccaya twice, once in June and July 1976, during the dry season, and again in January 1977, during the wet. That El Niño was a strange event that scientists would puzzle over for a very long time. On January 17, 1977, as Lonnie boarded a Braniff flight to Miami, the outside temperature in Columbus never edged above minus 9 degrees Fahrenheit. Somehow it seemed natural to think of searching for El Niño's fire in the layered ledgers of Quelccaya's ice.

FROM THE JAWS OF DEFEAT

Possibly the most remarkable thing about Lonnie Thompson is how unremarkable, at first, he seems. He is of medium height and medium build. His eyes are medium blue, and his hair is medium brown. His personality, too, seems painted in medium tones, inviting descriptions like "pleasant," "polite," and "nice." It doesn't take long, however, to realize that behind this deceptively placid façade lurks a man of imagination and ambition, determination and drive, especially drive. "Lonnie," says longtime friend Keith Mountain, "is relentless."

Keith Mountain well knows the man of whom he speaks. Together they have forded flooding Tibetan rivers on horseback, threaded their way through the treacherous serac fields of Peru's Cordillera Blanca, and endured temperatures of minus 30 degrees Fahrenheit in China's Kunlun Shan mountains. And over the austral winter of 1983 (south of the equator, winter starts in June rather than December), they spent ten brutal weeks at Quelccaya, pushing themselves well beyond what most people consider normal psychological and physiological limits.

Keith met Lonnie in 1978, soon after enrolling as a graduate student in Ohio State's geography department. Actually, he met Lonnie's wife, Ellen, first; at the time, she was a graduate student in the same department. A short time later, he and Ellen wandered over to the Institute of Polar Studies to find Lonnie. "Lonnie was stuck in this little office," Keith remembers, "and his lab was about the size of a telephone booth, and he would sit there in a silly-looking white hat and white coat, pulling ticker tape through his hands."

What Lonnie was doing, with the help of two machines known as Coulter counters, was methodically sizing and weighing minute particles of dust embedded, like microscopic splinters, in Quelccaya's mantle of snows. The job, Lonnie explained to Keith, had to be performed under ultraclean conditions. He was wearing that white hat and white coat to ensure that he did not contaminate the samples with the dust he carried around in his hair and clothes.

The Coulter counters spit out their measurements as a sequence of numbers on a steadily unfurling roll of tape, which Lonnie oftentimes would hand off to Ellen. First she would run the tape over to the electrical-engineering building, where it would be converted into a big stack of computer cards; then she would run the cards over to the computer center so that the data on them could be analyzed.

Lonnie started counting dust back in 1971, shortly after he showed up as a first-year graduate student in Ohio State's geology department. At the time, he was planning to become a coal geologist, largely because he had grown up in the state of West Virginia, where coal was king. Getting a good job ranked high on Lonnie's list of priorities. His father, he explained, had only an eighth-grade education and a bad heart, the result of a childhood bout with rheumatic fever. For as long as Lonnie could remember, the family had struggled to make ends meet. When his father died, Lonnie was an honors high school student in the little town of Gassaway, and to help out financially, he at one point held down four jobs. He then won a scholarship to Marshall University in nearby Huntington, which was where he met Ellen.

To Lonnie, coal geology seemed like a reasonable compromise between his intellectual interest in the earth sciences and the practical need to make a decent living. That was no doubt the reason that Lonnie at first expressed such disdain for glaciology. "When I was an undergraduate at Marshall," he smilingly recollects, "I can remember studying glaciers and thinking to myself, *What a waste.* Glaciers take up only a very small percentage of the surface of the earth; they're in really remote areas where people can care less about what happens, so why in the world would anyone take the time to bother to study them?"

But Lonnie also needed to earn his way through graduate school. So when a flyer appeared in his geology-department mailbox announcing that a research assistant's slot had opened up at the Institute of Polar Studies, he decided to apply. A short time later, he was looking at the dust particles entrained by one of the most famous lengths of ice scientists have ever recovered, the Byrd Station ice core from Antarctica. From top to bottom, it measured nearly 1.35 miles long and contained a continuous record of temperature and precipitation that spanned some ninety thousand years.

The Byrd Station core was pulled out of Antarctica in 1968, just two years after a U.S. Army Corps of Engineers drilling team retrieved a slightly shorter length core from the Greenland ice sheet that produced a climate record 140 thousand years long. From those cores scientists began to extract a high-resolution picture of the climatic conditions that prevailed during the last great glacial high stand, the one that commenced around 120 thousand years ago and peaked around 20 thousand years before the present.

What enchanted Lonnie and ultimately hooked him was the surprising eloquence of ice. To those who mastered its arcane language, he would discover, ice had so many stories to tell. Fluctuations in oxygen isotopes encoded swings between cold epochs and warm; fluctuations in nitrate levels encapsulated major expansions and contractions of vegetation; fluctuations in sulfate levels tracked volcanic explosions. Ice also contained organic material—from grains of pollen to winged insects—that could be used to reconstruct vanished ecologies. It entrapped air bubbles that could be used to reconstruct the composition of past atmosphere. And last but not least, ice recorded precipitation.

Precipitation could be directly deduced from the thickness of the annual layers of ice, after adjusting for compression, of course. It could also be extrapolated from the levels of dust in those layers. For despite its seemingly pristine whiteness, snow—and the ice that eventually forms from it—contains a surprising amount of dark detritus, from tiny dots of sand and flecks of dirt to the shelly carapaces of minute organisms. When the dustiness in a layer of ice rises, one can usually assume that prevailing weather patterns in a region are dry. When it falls, the converse is likely to be true.

The most important thing about dust, from a glaciologist's point of view, is that it blows on the wind, and it blows most copiously during intervals of dry weather. By comparison, during rainy spells, the dust swirling through the atmosphere is greatly reduced. The amount of dust in a layer of ice, then, provides what climatologists refer to as a "proxy" measure of precipitation. In principle, every swing between dry and wet—from seasonal to millennial—can be derived from feathery fluctuations in depositions of dust.

But that's not all. In places where the annual swings between wet and dry seasons are pronounced, the dust layers in an ice core form visible rings that can be used as dating tools. Dust rings, in other words, are not just proxy rain gauges; they are also proxy clocks that mark off time.

Lonnie's early work on dust swept him into a field that was exploding with excitement. In early 1974, right after he received his master's degree, he attended a workshop in Washington, D.C., attended by the Who's Who of the ice-coring world. At that meeting, Lonnie met Willi Dansgaard, who would turn into both an obliging collaborator and a formidable competitor. It was Dansgaard's lab in Copenhagen that performed the early oxygen-isotope analysis on the Quelccaya samples, at no charge to Lonnie. But it was also Dansgaard's lab that dealt Lonnie what, at the time, seemed like a tremendous blow.

Lonnie and Ellen had been hoping to do the dust analysis for a new Greenland ice-coring project that, in the early 1970s, was just getting under way. They had, in fact, submitted a proposal to the National Science Foundation asking for funding. As they soon learned, however, a researcher associated with Dansgaard's lab had his eyes on the same prize, and it was he who ended up doing that aspect of the analysis.

Lonnie was dismayed. The Greenland Ice Sheet Project, as it was called, was an international effort that involved the United States, Denmark, and Switzerland. But while both the Danes and the Swiss contributed a considerable amount, especially equipment and scientific expertise, the United States provided the largest share of support for the ice-coring operation. To Lonnie it seemed unfair that he would not be able to work on that ice. In his view, at least, he'd been frozen out of Greenland.

Today Dansgaard says that he is very sorry to hear that Lonnie felt that way, "and that he never told me about it," adding that if he had known, "We might have adopted him as a partner in our group, as we have done with other Americans." But it was just as well that Dansgaard did not adopt Lonnie, for it was the rejection of his proposal to work on Greenland's ice that caused him to turn to Quelccaya as a kind of last resort.

Almost immediately, in fact, Lonnie sketched out plans for another expedition and applied to the National Science Foundation's Office of Polar Studies for funding. In 1976, with a multiyear grant of over $100,000, he set off for Peru again. With him he brought five other hardy souls. Among them were geologist Alcides Ames, from the Instituto de Geologia y Mineria in Lima, and tropical meteorologist Stefan Hastenrath, from the University of Wisconsin in Madison.

As in 1974, Lonnie and his team set up camp between the twin moraines at Quelccaya's base. They remodeled the boulder-enclosed kitchen, expanding its size. Then they started to work. They dug more snow pits and retrieved more samples. They resurveyed the ice cap with a more accurate instrument. They hammered in two rows of stakes, perpendicular to each other, to measure the ice cap's growth due to snowfall versus its shrinkage from evaporation. And they placed a meteorological station on the top of the main summit dome, mounting it on fiberglass poles.

When they hiked out of Quelccaya in July 1976, they left the weather station on the summit dome, where it continued to record temperatures and wind speeds through December, at which point it was overturned by high winds and rendered inoperative. Lonnie might well have witnessed the storm that generated those winds, for he paid a brief visit to Quelccaya in January 1977, just to get a taste of what the weather in the wet season was like. Virtually every afternoon, he remembered, armadas of thunderheads menaced the ice cap, glowering down from heights of 70,000 feet. Compared to the clouds, Quelccaya looked vulnerable and small, and that greatly impressed him because, next to Quelccaya, he always felt vulnerable and small.

By this time, Lonnie was deep in the analysis of the meltwater the 1976 expedition had retrieved. He already knew, from his analysis of the samples he'd brought back two years earlier, that there was a decipherable climate record in the layers that made up the so-called firn, the term glaciologists use to describe snow that is more than one year old but has not yet turned to ice. But these samples came from only one site. Now that he had samples from several sites, he could finally select the best place to drill.

Lonnie went back to Quelccaya once more in 1977, during the dry season, and again in 1978, to lay the groundwork for an all-out assault. Up to then, the samples he'd obtained had been brought up by a heavy, barely portable drill that had to be operated by hand. It could not go much deeper than 36 or so feet. In 1979, however, he planned to use a helicopter to transport a polar-style drilling apparatus to the top of the ice cap along with the diesel fuel and generator needed to power it through to a depth of several hundred feet.

Lonnie, in other words, was determined to do in the tropics what, to that point, had been done only in Greenland and Antarctica—that is, he was determined to ream out a continuous length of ice that extended from the surface all the way to bedrock.

"One of the objectives of the ice analysis," he wrote in the funding proposal he submitted to the National Science Foundation, "is to establish whether an El Niño year leaves an identifiable record in the Quelccaya stratigraphy." Then he presciently added, "It would be scientifically very valuable to continue to collect meteorological data and annual snow collection on Quelccaya through a major El Niño period." The funding Lonnie asked for—about $130,000 spread over two years—came through, and in June 1979 he headed off for Peru again.

Everything very nearly came together the morning an excited crowd gathered at the edge of an open field in Sicuani. A Peruvian air force pilot had just landed in a twin-engine Bell 212 helicopter, and virtually everyone in town turned out to watch. Lonnie was counting on the helicopter to get the heavy equipment to the top of Quelccaya, but as soon as the pilot took a look at the load he'd been commissioned to carry—a cumbersome drill with 200-pound solid-steel cables, several 50-gallon drums of diesel oil, and a 435-pound generator—he suggested making a trial run. Lonnie, along with meteorologist Stefan Hastenrath, scrambled aboard, and the helicopter took off.

At first, the flight went smoothly; they had a panoramic view of the mountains and the ice cap ahead. But as the helicopter rose into progressively thinner air, the fuel-combustion process became more and more inefficient. Starved of oxygen, the engines choked to half-power. At that moment, the wind picked up, courtesy of a strong subtropical jet that sweeps across from the west at that time of year. As he struggled to maintain control of the chopper, the pilot abandoned all hope of landing. Instead, he headed back to Sicuani. There he informed Lonnie that he was backing out of the mission. It was too dangerous to approach the ice cap again, he said.

Lonnie stood there for a while, staring at his boots, his hands in the pockets of his down parka. Then he looked at the other members of his expedition and asked, with a wry smile, "Well, what do we do now?"

Together they mulled over the options. The fuel in the drums could be repackaged in smaller containers. With great difficulty, they might be able to break the drill into pieces small enough for horses to carry in and then for men to carry up. The showstopper was the generator. Not only did it weigh a quarter of a ton; it was also not easily disassembled. Even if they managed to get it apart, Lonnie observed, they probably wouldn't be able to get it back together again.

What they needed, mused University of Nebraska's Bruce Koci, head of the drilling team, was a completely different system, one that was designed with modular components that were meant to snap in and out. "Did such a system exist?" Lonnie wondered. "No," said Bruce, "but I think I can build one." A former aerospace engineer, Bruce was something of a technical wizard. He had, in fact, already designed and built a drill and used it to core ice at Canada's Baffin Island. To Lonnie, the idea that Bruce might be able to do the same for Quelccaya seemed like a long shot, maybe, but not preposterous.

The expedition continued. Leaving the unwieldy drill behind, they headed off for Quelccaya again. By the time they returned, Bruce Koci would be carrying a mental sketch of the drill he intended to build. It would be a thermal drill, he had decided, as opposed to an electromechanical one. That is, it would melt the ice around the core rather than cut it out. And it would be solar-powered. That idea occurred to him, he says, when he was standing on top of Quelccaya getting sunburned through industrial-strength sunblock. There was unlimited energy beaming down all around him, and unlike diesel fuel, it didn't weigh or cost anything at all.

After Lonnie got back from Quelccaya, he submitted another proposal for funding. This time the development of a portable, solar-powered drill formed the keystone of his plan. Willi Dansgaard was asked to review the proposal by the National Science Foundation, and Lonnie to this day delights in repeating the verdict he rendered. "Quelccaya is too high for humans, and the technology [to drill it] does not exist."

At the time, though, Dansgaard's comments were close to devastating. Over the years, Dansgaard had often been very helpful to Lonnie; in fact, he had offered to do the geochemical analysis of the ice cores from the abortive 1979 Quelccaya expedition at no charge. His expression of deep reservations at such a critical juncture Lonnie found deeply disturbing. "You know," Lonnie said to Bruce Koci in his measured, matter-of-fact way, "Willi says you can't do it." To which Bruce replied, "You can if you think you can." Lonnie laughed. It was the same motto he lived by.

The next years were spent developing the drill and the grid of solar panels needed to run it. The National Science Foundation, of course, wasn't about to fund Bruce Koci to develop a drill just for Quelccaya. It did, however, finance an effort to develop a drill for its polar programs. So with Quelccaya in mind, he designed a drill that could be taken apart and put back together with ease.

Lonnie, in the meantime, continued to return to Quelccaya to collect more samples. The most spectacular of these came from a 75-foot-deep crevasse that he and another member of the team rappelled into, only to find that the altitude had robbed them of the physical energy they needed to pull themselves up. Fortunately, the crevasse had a natural exit at one end, making it possible for the two scientists to lower themselves down to the bottom and then simply walk out.

The walls of the crevasse were as striated as a zebra's coat. Thanks to wind patterns that dominated the winter dry season, a gray line of dust separated each annual layer of snow. The lines were so clear that Lonnie was able to use them as a very easy way of assigning dates to the layers. The climate record in the wall of Quelccaya's crevasse, he determined, went back to 1962, which was an encouraging sign. It meant that there was a chance that Quelccaya's climate record might prove to be quite long; perhaps, Lonnie thought, long enough to track El Niño back in time for perhaps as much as 150 years. That estimate would turn out to be low by an order of magnitude.

By then, meteorologist Stefan Hastenrath had worked out a clear picture of the weather patterns that prevailed around Quelccaya and how El Niño affected them. The moisture that sustained the ice cap, he'd found, did not originate in the tropical Pacific, just several hundred miles distant, but in the Atlantic Ocean, almost 3,000 miles away. Clouds bearing that moisture rode the trade winds in from the ocean and across the Amazon. Along the way, these clouds continuously wrung themselves dry and then re-formed, so that by the time a snowflake fell on Quelccaya, the water it contained had been recycled who knows how many times by some Amazonian river or some leaf in the rain-forest canopy.

This east-to-west flow pattern was most pronounced from December through March, the summer wet season at this sub-equatorial latitude. Eighty percent of Quelccaya's annual 9–10 feet of snow fell at this time. During the winter dry season, the prevailing winds often came from the southwest, laden with dust swept from the bare fields of the Altiplano, where, in addition to potatoes, the crops include other indigenous tubers, like oca, and grains, like quinoa and cañihua. Yet even then, the moisture Quelccaya captured in the form of snow usually came from the east, from the Amazon.

During El Niño years, the flow of moisture from the Amazon

to Quelccaya slowed. In direct contrast to the northern desert around Paita, El Niño's signature at Quelccaya was not rain but drought. In the Quelccaya crevasse, for example, the El Niño's of 1972–73 and 1976–77 showed up with startling clarity as reductions in the thickness of the annual layers of snow. And because of the measurements Lonnie had so carefully taken in the years before and after, these layers could be directly compared to those laid down in so-called normal years. When converted into rainfall equivalents, these measurements of snowfall indicated that in El Niño years precipitation dropped by a rather large 30 percent.

The results from the crevasse appeared in a paper published by the journal *Science* in 1984, a year after an extraordinarily powerful El Niño had torn through the world. But while that paper generated a flurry of interest, it was nothing compared to the excitement that greeted the publication of the paper that appeared the following year. That was the landmark paper, the one that launched Lonnie's career. It showcased the analysis of the ice cores Lonnie had retrieved in 1983, on an expedition few thought would succeed.

Lonnie started gathering together the equipment for that expedition in the spring of 1983, all the while holding his breath. In May, he and Bruce Koci tested the new drilling apparatus on a big piece of ice in a campus parking lot. Bruce remembers that some students passing by wondered if he was creating a piece of ice sculpture. The drill worked, or at least it worked on a block of ice that was about 3 feet thick.

The two men talked about what they would do if the drill failed. Lonnie said that he'd already decided. This time, if he came back from Quelccaya empty-handed, he was going to quit glaciology for good. He had taken the entrance exam for Ohio State's MBA program, he told Bruce. If he couldn't do science, he said, he might as well make some money.

In June, Lonnie and his small expeditionary force walked into Quelccaya behind a caravan of some forty horses (with a few mules and donkeys thrown in for good measure). It was a colorful sight. The horses were roans and blacks and grays with glossy coats, and the bags that contained the solar panels were bright red. Bruce Koci had specially designed them to protect the panels; horses, he'd noted on a previous expedition, had a favorite ploy for ridding themselves of loads. They rolled over on them.

At that point, the National Science Foundation had still not come through with the funding. To cover expenses, Lonnie had cajoled an advance out of Ohio State's Institute of Polar Studies. Bruce Koci, for his part, had convinced the University of Nebraska to ante up the $40,000 needed to purchase a 2-kilowatt array of solar panels.

In June 1983, the historic El Niño that had begun the year before had still not fully abated. Floods and landslides had trashed the north coast of Peru, melting many of the lovely adobe friezes at Chan Chan, and severe drought had descended on the Altiplano, decimating the potato crop. Given what he knew of El Niño's impacts, Lonnie was expecting Quelccaya to be dry. Even so, he was not at all prepared for the sight that awaited him. For the first time ever, Quelccaya looked dirty to Lonnie, its snowy outer garment peppered with grit and grime. Seventeen years from now, he reflected, when the snow that fell in 1982–83 completed its transition from firn to ice, that grit and grime would stand out as an extremely thick band of dust in an unusually thin layer of ice.

Lonnie and the others unloaded the horses; they set up camp again in the usual place, between the two moraines. The next day, they turned into beasts of burden and started hauling equipment to the top. At first, they could carry only two solar panels; as they became acclimatized, they added a third. Even so, it took the better part of a week to get all the panels into position. At that point, they started hooking up the solar array. There were around fifty panels in all, with about a dozen spares, and they worked surprisingly well, so well that touching one was guaranteed to deliver a nasty electrical shock.

Then, at a site near the summit, Bruce Koci started to drill. At first, it didn't go well. The top layers of the firn resisted the drill bit; the core came up in little chewed-over chunks. Then, right around the interface between the ice and the firn, the drill bit hit the substance Lonnie most dreaded finding: water. The upper portions of the ice cap, Lonnie knew, were almost exactly 32 degrees Fahrenheit, which meant they were right on the edge of melting.

As Lonnie held his breath, Bruce Koci started in again. The drill's heated bit traced out a cylindrical path, cutting through the ice the way a circular saw slices through wood. After a time, Bruce stopped, pulled the drill pipe up, and shook the core out. "You should have seen the look on Lonnie's face as that core came out," Koci said, laughing. "His eyes got bigger as the core got longer. He knew then that this was going to work."

Electromechanical drills bring up cores that are cloudy looking, occluded by ice shavings. Bruce's thermal drill brought up a polished icicle with the clarity of a diamond. This icicle, though, was 6 feet long; clearly visible were the dust bands that marked the annual dry seasons. That's when Lonnie knew: The water that coursed through the upper layers of ice had not erased the climate record in the ice below.

They settled into a rhythm. Every day three or four of them would trudge up from the base camp, and every day Bruce Koci would drill out around 40 feet of core. Someone, often Lonnie, would sit behind a little snow wall that acted as a windbreak and measure the core, photograph it, assign it a number, and in a waterproof field book, write down its distinguishing features. Using the dust rings as markers of the annual layers, he would then take out a little saw and start cutting the core into chunks that represented a year's worth of snowfall.

The chunks of ice were placed in leakproof plastic bags, each clearly marked with a numerical code that indicated its vertical position in the core. As soon as there were enough chunks of ice to fill a backpack, one member of the team or another would transport the haul down to a special processing tent that had been set up at the camp. The tent served to trap the sun's heat so that the ice would melt more quickly. In the end, Lonnie brought back some six thousand sturdy plastic bottles filled with Quelccaya's meltwater.

The work was nonstop, and at that elevation, it was also brutal. The ultradry air robbed moisture from every body surface it touched. Unfiltered by clouds, the sun rained down shards of ultraviolet radiation. Skin burned and bled and cracked and chafed. Blisters and sores erupted on feet and faces. At one point, Keith Mountain remembers, Lonnie's bottom lip ballooned to twice its normal size. It drooped into his soup, his oatmeal, whatever he was eating. He didn't complain, and neither did Keith, who by now was so badly sun scorched that, in desperation, he went around with a big patch of duct tape plastered across his nose.

They soldiered on. After Bruce pulled up the last length of core, he said to Lonnie, or maybe Lonnie said to him, "I don't think one core is adequate. I think we need to do another." By now they were running low on food. Keith Mountain went out to Sicuani to bring back more supplies. While there, he sent back word that the expedition had succeeded beyond anyone's wildest expectations. A short time later, the National Science Foundation announced it was coming through with the long-awaited funding.

By that point, Bruce had started in on the second drilling site, which lay on the summit itself. Very late in the afternoon, the day before they were scheduled to leave, he passed the 450-foot mark. Shortly thereafter, the drill hit what is known as an unconformity, a place where the integrity of the ice broke down. At that point, Lonnie declared the expedition over. The sun was about to set; the horses were due in early the next morning, and before then, everything on top of the ice cap had to come down. Hastily, he and Bruce and the others piled everything on top of the drill's winch and, using the winch as a sled, ran with it downhill. They made it to the edge of the ice with minutes of daylight to spare.

The next morning, gaunt, haggard, unkempt, they broke camp and began the long trudge out. Almost beyond exhaustion, Lonnie drifted in and out of a dreamlike euphoria. He was coming back with the meltwater from not one but two ice cores. The first core was more than 480 feet long; the second, more than 460. And those cores—Lonnie already knew because he'd counted the dust rings—encapsulated a climate record that spanned fifteen hundred years.

Yet even Lonnie did not suspect the richness of the climate record that would eventually emerge. The Quelccaya Ice Cap was like a weather station that had been in continuous operation since A.D. 500, more than a millennium before the Spanish arrived. In its layers archaeologists would find evidence of the El Niño floods that were said to have devastated the great coastal city of Chan Chan around A.D. 1100; they would find evidence of the decades-long drought that decimated the pre-Incan civilization of Tiwanaku at around the same time. And they would also find evidence of centuries-long El Niño– and La Niña–like periods that eerily paralleled the rise and fall of these and other pre-Incan cultures.

But on the day he left Quelccaya in August 1983, Lonnie was not looking forward so much as back. During a quiet moment, he scribbled a last entry in his orange-jacketed field book. It was a simple note of thanksgiving to Bruce Koci, Keith Mountain, and all the others who had stuck by him for so long.

CHAPTER **FOUR**

Canst thou draw out Leviathan with a hook?

Book of Job

SURPRISE

For what seemed like hours, Ants Leetmaa wandered through the redolent aisles of Zanzibar's Darajani Market. Along rows of rough-hewn tables fishmongers hawked shiny sardines and steaks cut from fresh-caught marlin, and fruit vendors displayed mangoes and jackfruit, bananas and pineapples, in handwoven baskets. All around him the air hummed with the buzz of flies and the drone of conversation. At last, Ants stumbled across a narrow alleyway that served as the spice section. There, amid colorful mounds of turmeric and saffron, he found what he was looking for, piles of cloves whose pungent aroma enveloped passersby in exotic perfume. He bought several bags to bring home as souvenirs, then made his way back down to the harbor, relieved to be out in the open again.

The year was 1970. Ants had just received his Ph.D. in oceanography from the Massachusetts Institute of Technology (M.I.T.) and almost immediately had set off for the Indian Ocean to study the striking shift in currents that occurred twice a year off the East African coast. In Zanzibar he chartered a vintage pleasure boat of questionable seaworthiness and a crew of slightly shady characters, then continued north to the Kenyan port of Mombasa. From there he ventured out every two weeks to monitor the spin-up of the Somali Current, whose dramatic swings from activity to inactivity made it seem like a manic-depressive version of the Gulf Stream. At the time, virtually nothing was known about the Somali Current, and that greatly appealed to Ants.

As an undergraduate at the University of Chicago, Ants had played around with the idea of becoming a particle physicist, but in his senior year, he chanced across a slender book in a light green cover that caused him to change course as abruptly as the Somali Current itself. The book was *The Gulf Stream* by Henry Stommel, and it began with an account of the explorer Ponce de León, who, in the year 1513, Stommel explained, "crossed the stream north of Cape Canaveral and then sailed south to Tortugas. The current was so swift that his three ships were frequently unable to stem it."

Ants was immediately captivated, not just by the romance of the river in the ocean but also by Stommel's use of observational data to challenge cherished beliefs. "There is scarcely any more firmly rooted idea in the mind of the layman than the notion that the Gulf Stream keeps the European climate warm," Stommel wrote. "It now seems that it is not so much the Stream itself that is important, as the position and temperature of the large mass of warm water on its right-hand flank." Almost before he finished reading the book, Ants found that he had reached a decision. He would go to M.I.T. and study with Stommel.

Stommel, Ants soon discovered, had a mind of Olympian sweep, combining a theorist's intuitive feel for the flow of fluids with an experimentalist's knack for grounding theoretical musings in hard data. "Why do our ideas about the ocean circulation have such a peculiarly dream-like quality?" Stommel asked in a famous paper, then went on to answer the question he'd posed. If oceanographic theories too often resembled ships cast adrift, Stommel observed, it was because they were unanchored by physical evidence. "At the time," Ants later reflected, "we knew almost nothing about the ocean."

But talking about physical evidence as grist for theory was one thing; gathering it, quite another. This was something Ants appreciated as soon as he ventured out into the Indian Ocean. For starters, the Somali Current proved to be exceptionally strong— "up to five or six knots, which is eight miles an hour," Ants reminisced, "and when you're out there, you realize that there's nothing between you and India." Add to that "an old clunker of a boat, a captain who was a high school dropout from England, guys on the crew who still believed in witch doctors, and a second officer who was a South African trained by Russians" and Ants realized that he had inadvertently concocted a recipe for trouble.

"So now you've got all these chronically unhappy people on board, and you're out there, and the current is beginning to surge, and the chief engineer comes up to say that one engine is down, and he's not sure there's enough fuel to get back to Mombasa. The boat always managed to squeak back in, but it was a little nerve-racking."

In other ways, though, it was a paradisiacal time.

In between voyages, Ants spent many pleasurable hours exploring islands with white-sand beaches and tumbledown ruins littered with shards of blue-and-white Chinese pottery. His base was the city of Mombasa, where the port was thronged with dhows, traditional sailing vessels that have plied the Indian Ocean for more than two thousand years. They were sturdy craft with wide triangular sails that could be drawn close to the wind or let out in front like a spinnaker. Many sported carved prows painted red, white, and green, and on either side were circles containing a star and a crescent; "dhow's eyes," locals called them.

These were the ships that had opened up lucrative trade routes to Arabia, Persia, India, and beyond, thereby sustaining a string of prosperous sultanates that flourished along the East African coast from the thirteenth century onward. India proved to be a particularly important link in the chain. In India, dhows from Zanzibar, Mombasa, Malindi, and Lamu took on Chinese porcelain and silk; their Chinese counterparts loaded up with gold and ivory and rhinoceros horn brought out of the African interior. A Ming dynasty fleet once brought back from Bengal a giraffe that had been shipped from East Africa by dhow.

At the end of the fifteenth century, a fleet under the command of Portuguese explorer Vasco da Gama rounded the Cape of Good Hope and made its way north to Malindi. There they hired a navigator who was familiar with the currents and the winds of the monsoon system, and the following month the fleet sailed into Calcutta.

What enabled the dhows to traverse such great distances was the seasonal shift in winds and currents that characterizes the East African monsoon. The northeasterly winds that blow from October to March brought the dhows south to their home ports of call; the southwesterlies that blow from April to September sped them north again. In the first instance, the dhows were assisted by a slow south-moving current; in the second, by a fast-moving northward flow—the Somali Current—that turned on in the spring and shut down in the fall.

As its name suggested, the Somali Current was particularly pronounced off the coast of Somalia, whence it surged north for nearly a thousand miles. The question was, did the shift in the winds drive the shift in the currents, or was it the other way around? In an attempt to provide an answer, Ants and his crew methodically flung out dozens of buoys into the ocean as they approached the Somali border. Somalia had recently turned Communist and was virulently anti-American, so they always turned back once they ventured close to the border, relying on radar to track the buoys.

For the first few voyages, the buoys floated slowly south. Then, around the third week of April, the Somali Current abruptly turned on, and the buoys started moving swiftly to the north. The switch in the current, Ants established, neither led nor lagged the shift in the winds, but appeared in near-flawless synchrony. In this respect, the monsoon regime in the Indian Ocean bore a strong similarity to El Niño, though at the time Ants was not thinking about El Niño any more than Lonnie Thompson was thinking about El Niño prior to his first foray to Quelccaya. He was simply doing very basic research in a region of the world that scientists in his field had barely begun to explore.

At the time, most oceanographers harbored the belief that currents circulating through the oceans were quasi-permanent features and varied very little on time scales of less than a thousand years. The Somali Current stood out as an intriguing exception. Just as it seemed to Lonnie Thompson that a tropical ice cap might offer clues to the ways in which the atmosphere responds to climatological change, so it seemed to Ants that a tropical current might provide the keys to unlocking an equivalently deep mystery surrounding the ocean.

Shortly after finishing his study of the Somali Current, Ants took a job as a research scientist at the Atlantic Oceanographic and Meteorological Laboratory in Miami. In 1975 and 1976, he made two more forays to the Indian Ocean, this time to study the equatorial undercurrent, a subsurface leak-back of water piled up by the winds at one end. Again, his objective was simply to map the current: its depth, its velocity, how it changed from one season to the next.

As his base, Ants chose the Seychelles, with their beautiful white-sand beaches and huge granite boulders that seemed to have somersaulted down to the sea. He hired a boat that, from afar at least, looked like a classy yacht but on closer inspection revealed numerous flaws. For one thing, the boat turned out to be cockroach infested. For another, it carried loose rocks in its bottom as ballast, and so "if the boat started to go," Ants reflected, "you'd be gone, too."

From then on, Ants shifted his attention to the larger bodies of water that bordered the American coastline, the Atlantic and the Pacific. Increasingly, he found himself going to sea in big ships that sat high off the water and were expensive to run, so that casual stops to explore a nearby island or take an afternoon swim were not so easy to arrange. After a while, the tedium of the voyages began to weigh on him, as did the feeling of treading water.

"At first, almost anything you saw was something new," he remembered, "and there was no problem that was too small to think about. But then you started seeing the same things over and over again."

Ants started dreaming of change. His wife had a job offer in Walnut Creek, California. "I thought maybe I could go to Walnut Creek and become a walnut farmer," he said only half in jest. He also thought of going to Rhode Island to raise chickens, Rhode Island Reds, of course. Then again, he thought he might indulge a suppressed artistic bent and become a painter or a writer. Somewhere in the midst of this existential crisis, he divorced and remarried.

Intellectually, the turning point came in November 1982, when Ants found himself on a ship named the *Researcher* as it crisscrossed the eastern basin of the equatorial Pacific between the South American coastline and the Galápagos Islands. He could immediately see that the ocean had undergone some radical change. Earlier in the year, phytoplankton had been so numerous that they dyed the ocean a goldish shade of green, and when biologists strained the water through filters, they picked up an aroma that reminded them of fresh-cut hay. Now, all around the ship, the water had turned blue, the color that signified an oceanic desert.

Beneath the surface, things were different, too. This time, when he used his instruments to probe the water, Ants could see that the whole physical structure of the ocean had warped somehow, like a length of wood left too long in the rain. Cold water that used to lie close to the surface had plunged to unheard-of depths. Currents that were supposed to travel west along the equator appeared to have reversed course, so that they now ran east. Plus the thin rivulet known as the equatorial undercurrent had vanished. It had stopped flowing as suddenly and inexplicably as the winter creek whose ominous blockage, some fifteen years later, would presage the Rio Nido, California, landslide.

Ants had just turned forty years old. The biggest El Niño scientists had ever witnessed was ramping up to full power, and waters that had seemed so familiar a short time before had turned exceedingly strange. To Ants, the tropical Pacific suddenly seemed a lot more interesting than it had in a very long time.

The *Researcher* ventured into the tropical Pacific again in March 1983 and once more in July, by which point the ocean's structure had very nearly returned to its normal state. It was a flip-flop even more dramatic than the reversal of currents in the Indian Ocean, and this time, Ants realized, oceanographers were faced with a very different problem than the one that had so troubled Henry Stommel. Over the years, Ants and his peers had managed to amass quite a bit of data. Now they had to figure out ways to put those data to use.

WHILE SCIENTIFIC PROGRESS may be measured by its successes, it is urged forward by its failures, and it was the failure of scientists to spot the onset of the 1982–83 El Niño that provided the emotional and intellectual energy that powered the next series of advances. This was not just any El Niño that the scientific community missed, emphasizes Florida State University meteorologist James O'Brien, but an El Niño of historic proportions. O'Brien, for one, has yet to get over the sense of shock he experienced at the time. "It was the biggest El Niño of the century, and it had almost peaked before we even knew it was happening!" he still wonderingly exclaims.

Today it seems unthinkable that such a monstrous climatological upheaval could have caught the scientific community unawares, but that's exactly what happened—and the comedy of errors that led to this oversight started with the eruption of an obscure volcano in southern Mexico by the name of El Chichón.

In the early spring of 1982, El Chichón was just a little lump of rock that didn't even measure a mile high. Its forested slopes sheltered a few fumaroles that belched from time to time, but the Zoque Indians who lived around the caldera could not remember a time when the volcano had seemed truly menacing. Suddenly, however, this diminutive cone shell of a mountain started spewing forth plumes of sulfurous steam. It also unleashed a series of tremors.

Then, on March 29, as a swarm of earthquakes rumbled a warning, El Chichón exploded. A terrible blast of superheated rock spread out at supersonic speeds, scouring the surrounding terrain, gouging channels into bedrock and stripping the bark from broken trees. Over the coming days, more explosions followed. As the searing heat caused rivers to boil and incandescent ash set houses ablaze, thousands of people fled in terror, and casualties quickly climbed into the hundreds. In all, El Chichón destroyed nine villages and claimed over two thousand lives.

Even today the power of the 1982 eruption remains evident in the contours and even the colors of the landscape. El Chichón used to be 4,134 feet tall. It currently stands just under 3,500 feet, and instead of a roundish dome, it is now capped by a caldera that measures close to a mile across. Heavily tainted by acidic ash, the lake at the center of the caldera gleams an unearthly aquamarine.

In all, El Chichón threw off about the same amount of ash as Mount St. Helens did in 1980—enough to raise a football field 60 miles high. But whereas Mount St. Helens jettisoned most of its debris near ground level, El Chichón aimed at the skies. As a result, the total amount of material that El Chichón lobbed into the upper reaches of the atmosphere was at least ten times greater. Moreover, the stuff El Chichón coughed up turned out to be inordinately rich in sulfurous gases. Combined with water vapor, these gases very quickly seeded the stratosphere with minuscule droplets of sulfuric acid, the most important of the volcanic aerosols.

Quite simply, El Chichón was the first climatologically important volcanic eruption to occur in the satellite era. In fact, there hadn't been anything like it since 1912, when Alaska's Katmai erupted in similar rocket-launcher style. This time, though, scientists were able to track the movements of voluminous clouds of dust and ash and gas hurled by successive eruptions high into the stratosphere. They could also measure the changing contents of the clouds as the heavier particles fell out and the lightweight aerosols just hung there and hung there and hung there.

Scientists had known about the connection between volcanoes and climate for two centuries, and the explosion of El Chichón immediately evoked predictions that the earth would be measurably cooler than usual over the following year. "As far back as 1783," observed *TIME* magazine, "when the eruption of Iceland's Laki volcano was followed by an extremely cold winter, Benjamin Franklin pointed out that volcanic dust could partly block sunlight. After the explosion of Indonesia's Tambora volcano (in 1815), there were so many frosty nights . . . in Canada and the U.S. that people called it the 'year without a summer.' "

Tambora, it should be noted, was the largest volcanic eruption in something like ten thousand years, and as with El Chichón, the aerosols, not the dust, had been the major culprits. Long after Tambora's dust settled, these featherweight droplets remained; they cooled the surface of the earth by scattering solar radiation and reflecting it out to space. Indeed, the aerosols lingered in the atmosphere for so long that their cooling effects were felt far into the following year, when Jack Frost made sporadic visits to New England and Canada from late spring into early fall. "Melancholy weather," moaned the Danville, Vermont, *North Star*. Across the Atlantic, in the British Isles, summertime temperatures dropped below normal by as much as 5 degrees Fahrenheit. Lord Byron reflected the broadcast gloom in a poem he titled "Darkness." "Morn," he wrote, "came and went—and came / and brought no day."

For months, the flamboyant eruption of El Chichón dominated the pages of scientific journals and popular newspapers and magazines. "El Chichón Forebodes Climate Change," worried the journal *Science*. "Volcanic Cloud May Alter Earth's Climate," fretted the *National Geographic*. *Weatherwise* went so far as to wonder if El Chichón's dramatic atmospherics might qualify it for the title of "Weather Maker of the Century." In fact, El Chichón's high clouds only temporarily overshadowed the nearly simultaneous emergence of the 1982–83 El Niño, a weather-maker that would prove to be more powerful by far.

HAD IT NOT BEEN for the eruption of El Chichón, it is often said, scientists *might* have realized that a big El Niño was in progress

much earlier than they did. And certainly the reflective veil of sulfuric acid droplets in the stratosphere compromised the ability of
satellites in low-earth orbit to render accurate readings of sea-
surface temperatures. Like eyes peering through a pair of foggy
lenses, the instruments mounted on these satellites squinted
mightily at the tropical Pacific, then made it out to be as much as
6 degrees centigrade, 10 degrees Fahrenheit, colder than, in fact,
was the case.

But the satellite readings were hardly the only source of mis-
information, according to meteorologist Eugene Rasmusson,
who, at the time, was one of the branch chiefs at the National
Weather Service's Climate Analysis Center in Camp Springs,
Maryland. After all, as he later put it, "we still had surface observations from ships." In principle, the observations from the
ships could have been used to adjust the satellite data, which everyone realized had to be skewed. The trouble was, as Rasmusson
and others were chagrined to discover later on, the ship data to
which they had been paying attention were also wrong.

In retrospect, it was an embarrassing mistake. It was as if min-
ers panning for gold in a river had thrown out all the nuggets that
were shiny and yellow and kept the drab bits of gravel instead.
How could this happen? Well, measurement may be the core of
science, but as scientists know, it is also subject to all sorts of biases, and the way to correct for one common source of bias is to
throw out readings that seem off-scale. Prior to 1944, for example, merchant seamen used to measure sea-surface temperatures
by scooping a bucket of water out of top layers of the ocean.
Many of them were extremely meticulous and took their readings
right away, but others would become distracted by some more urgent task. As a result, the bucket of water would sometimes sit
too long on deck and heat up in the sun, producing bogus temperatures.

Doing away with known biases, of course, is easy, but doing
away with those that scientists themselves inadvertently introduce
is far trickier. In 1982, it was the system of quality control scientists imposed on their data that created the problem. As Rasmus-

son recalled, the computer program at the Climate Analysis Center that processed sea-surface temperatures had thresholds requiring it to throw out readings from merchant ships (ships of opportunity, they were called) when they exceeded a certain temperature range.

Establishing these thresholds had not been a capricious step; everyone knew that seamen could be sloppy, that instruments on ships sometimes malfunctioned, that land temperatures confused coastal readings. The problem was, the thresholds had been set artificially low because no one had imagined that ocean temperatures could ever rise as high as they did in 1982 and 1983—that they could come to resemble temperatures in a bucket of water that some careless seaman had let stand too long in the sun.

In truth, though, it was not just the lack of reliable data that blinded the scientists to the power of the El Niño developing along the equator. An even greater problem, reflected Rasmusson, was that he and many others were carrying around in their heads too pat a picture of the way El Niños *should* behave.

Just the year before, Rasmusson and his colleague Thomas Carpenter had published a now-famous paper describing what came to be known as the "canonical" El Niño. From six El Niños that occurred between 1949 and 1976, they concocted a composite portrait, rather like police artists creating a sketch of a suspect from a half-dozen eyewitness descriptions. El Niño, they said, had the following modus operandi: It typically started early in the year, during the late northern winter or early spring, and it appeared first along the South American coast, then spread into the central Pacific. In 1982, however, nature laughed at Rasmusson and Carpenter. Not only did the new El Niño fail to show up on cue, but it also recited its lines in reverse, so that the waters of the central Pacific warmed months in advance of the coast.

And that's not all it did "wrong." The El Niño of 1982–83 also nonchalantly skipped over what many scientists had come to consider an essential precursor phase, as vital to the subsequent drama as the scene between Macbeth and the witches or Hamlet and his father's ghost. Influential University of Hawaii oceanographer Klaus Wyrtki had dubbed this phase the "buildup," and central to it was the dialogue imagined by Jacob Bjerknes, so that changes in the ocean provoked the atmosphere and changes in the atmosphere provoked the ocean.

According to Wyrtki, the buildup worked like this: First, an increase of sea-surface temperatures in the western Pacific served to intensify the process of convection, that is, the rate at which air warmed, rose, and ballooned into clouds. The intensification of convection then triggered a commensurate drop in atmospheric pressure, which deepened the low in the vicinity of Indonesia. The deepening low, in turn, caused the trade winds to blow faster, and the faster they blew, the more water they swept from the east and the higher they piled it in the west. Over time, it was easy to imagine, the western side of the ocean would become more and more steeply sloped and therefore more and more unstable, like an alpine snowfield in advance of an avalanche.

A buildup, Wyrtki had determined, had preceded the El Niños of 1972–73 and also 1976–77. In early 1982, however, there was no evidence that a buildup had commenced, and in June he confidently declared that there would be no El Niño that year or the following year. In September and again in October, he repeated the assertion. By that point, however, others were starting to pick up on signs that, buildup or not, a new El Niño had begun.

Among the first to sound the alarm was a team of Peruvian fisheries researchers who'd been hired by an international project to take daily readings of sea-surface temperatures. They were based in the port city of Paita, and always they followed the same routine. First they walked to the end of a rickety wooden pier and measured the temperature of the water in the harbor. Then they climbed into a diesel-powered fishing boat providentially named *El Niño del Milagro* (in Spanish, *milagro* means "miracle") and chugged out into the open ocean. There they repeated the same measurements.

Between September 22 and September 23, 1982, the Peruvians caught a rise of nearly 4 degrees Celsius, 7.2 degrees Fahrenheit, in a single day. This startling uptick was relayed to U.S. scientists in early October when one member of the team stood at a pay phone in Paita and read off a month's worth of data to Francisco Chavez, who was then a graduate student in biological oceanography at Duke University. Neither Chavez nor his Ph.D. adviser, Richard Barber, immediately appreciated the significance of the rise; in fact, they thought it was a fluke fluctuation that would soon disappear.

A short time later, scientists from the Woods Hole Oceanographic Institution ventured into the eastern Pacific on the research vessel *Conrad* and started radioing back measurements of temperatures, currents, and winds that were equally astounding. Their transmissions catalyzed an intense debate over what was happening in the tropical Pacific. The argument was finally settled in November when the *Researcher*, the ship that Ants Leetmaa was on, began its own systematic sweep of equatorial waters.

What was tantalizing about the fall of 1982 was that so many scientists had an inkling that something was amiss but couldn't quite pinpoint what it was. On September 22, for example, the day before the Paita biologists recorded that remarkable temperature rise, Eugene Rasmusson and his colleagues at the Climate Analysis Center issued a special bulletin entitled "The Global Climate Fluctuation of June–August 1982." In that bulletin they noted:

> One of the most significant climate anomalies of the past season has been the persistent and locally severe drought conditions over the Indo-Australian region. . . . The global pattern . . . has many of the characteristics of a climate fluctuation generally referred to as the Southern Oscillation. First discovered by the British climatologist Sir Gilbert Walker about 50 years ago . . . it has become the subject of renewed interest during the past several years as its relationship to large-scale ocean temperature changes in the eastern equatorial Pacific and climate fluctuations in middle and high latitudes become apparent.

On November 10, 1982, Rasmusson's group followed up with a second bulletin. This time they made direct reference to a "warm event," an El Niño, that is. Over the past six weeks, they reported, the drought in the Indo-Australian region had intensified. Plus sea levels had risen in the central Pacific to such an extent that some of the Line Islands—Christmas Island (now known as Kiritimati) in particular—were experiencing floods in areas "that have not been flooded in the memory of the inhabitants."

On February 14, 1983, the Climate Analysis Center put out a bulletin that took note of worsening conditions in parts of the Indonesian archipelago, where "one million people . . . face famine because of crop losses during the drought." By then, the Andean country of Ecuador had suffered such terrible rains that another million people, "one-eighth of the country's population," noted the Climate Analysis Center, "were at risk for cholera and other diseases that spread in the wake of widespread flooding."

On April 15, 1983, the Climate Analysis Center issued yet another update, noting the coincidence of unusual storminess over French Polynesia, with exceptionally heavy precipitation across the southern tier of the United States, from California all the way to Florida. Terms like *unparalleled, highly anomalous,* and *record* peppered the pages of this otherwise tersely worded report.

Like an ambush attack, the El Niño of 1982–83 revealed gaping holes not only in the theoretical ramparts but also in observational capabilities. Suddenly, monitoring the state of the tropical Pacific seemed far too important to be left to a patchwork of weather satellites, occasional research vessels, and meandering merchant ships, and over the next few years, scientists drew up ambitious plans to rectify the situation. Among other things, they designed a network of instrumented buoys, which they then proceeded to string across thousands of miles of equatorial ocean. They also designed dedicated satellite-mounted instruments for monitoring the sea's surface on a real-time basis.

Very soon, Ants Leetmaa realized, billions of bits of data—far more data than he had ever collected from the deck of a research vessel—would be flying out of the tropical Pacific. As he saw it, the challenge no longer lay in collecting information about the ocean but in making use of it. Only a very large computer model, he reasoned, would be able to sift through the extravagant amount of information that was about to pour forth and separate from the chaff the kernels of meaning.

WYRTKI'S WAVES _____

The Pacific Ocean is the earth's most conspicuous feature. It extends from the frozen continent of Antarctica in the south all the way to the Bering Straits in the north, and at its widest point, just to the north of the equator, it stretches more than halfway around the world. The first European to lay eyes on its shimmering waters was Vasco Nuñes de Balboa, who, in 1513, waded into the surf off the west coast of Panama, sword held high, and claimed for Spain the unexplored vastness beyond the horizon.

Yet despite his keen sense of drama, Balboa could not have guessed the immensity of his discovery. The waters of the Pacific cover fully a third of the earth's surface, a staggering area of 63,800,000 square miles. All seven continents could fit inside its expansive boundaries, and the Atlantic and Indian Oceans as well. On average, the Pacific is around 14,000 feet deep, and east of the Philippines, in the Mariana Trench, it drops away to the dizzying depth of 36,000 feet. At that point, the Pacific is deeper than Mount Everest is tall.

The Pacific, Melville wrote, "rolls the mid-most waters of the world, the Indian Ocean and Atlantic being but its arms. The same waves wash the moles of the new-built California towns, but yesterday planted by the recentest race of men, and lave the faded but still gorgeous skirts of Asiatic lands . . . while all between float milky-ways of coral isles, and low-lying, endless, unknown Archipelagoes, and impenetrable Japans. Thus this mysterious, divine Pacific zones the world's whole bulk about; makes all coasts one bay to it; seems the tide-beating heart of earth."

But while El Niño influences this huge expanse of water and is influenced by it in turn, its main theater of action is confined to the sparkling, sunlit sash that girdles its midriff—the equator.

"It looked like a blue ribbon stretched across the ocean," Mark Twain wrote of the world's most intangible landmark as he crossed it on a passenger ship steaming down from Hawaii. "Several passengers kodak'd it. We had no fool ceremonies, no fantastics, no horseplay. . . . In old times, a sailor, dressed as Neptune, used to come in over the bows, with his suite, and lather up and shave everybody who was crossing the equator for the first time, and then cleanse these unfortunates by swinging them from the yard-arm and ducking them three times in the sea. This was considered funny. Nobody knows why." In contrast to Twain, Melville found no need for tongue-in-cheek contrivance where the equator was concerned. Like other seamen, he simply referred to it as the Line.

Geographically speaking, the equator *is* just a line, an imaginary line that wraps clear around the world. Geophysically, however, it is a real, and rather strange, place, courtesy of the so-called Coriolis force.

The Coriolis force arises as a consequence of the earth's rotation about its axis, and it bears the name it does because its importance was first noted by Gaspard G. Coriolis, a French mathematician and engineer who lived in the nineteenth century. Unlike gravity, the Coriolis force exists only within a rotating reference frame, the reason that it is sometimes referred to as an "apparent" force.

To explain how the Coriolis force operates, scientists often invoke the image of a spinning turntable and invite one to imagine trying to roll a bowling ball across it, draw a straight line on it, or pitch a baseball to a batter standing on the rim. Invariably, the bowling ball will miss the pins, the line will curve, and the baseball will never enter the strike zone, unless, of course, the bowler, the drawer, the pitcher, adjust by exerting a countervailing force of their own.

In one of his famous lectures at the California Institute of Technology, the late physicist Richard Feynman tried to give students an intuitive feel for the Coriolis force by telling a little story about a character named Moe who kept trying to push an object from the center of a moving carousel out to its edge. With each step he took, Moe discovered, the more torque—that is, sidewise force— he had to exert in order to keep the object from flying off to one side. The most important attribute of the Coriolis force, Feynman explained, is this: An object experiences it most strongly as it moves along a line that is perpendicular to the axis of rotation, that is, along the carousel's radius. "To walk along the radius in a carousel, one has to lean over and push sidewise. Try it some-time," Feynman told his students.

The reason is not hard to understand. A carousel, like an old-fashioned hi-fi system's turntable, spins around at some number of revolutions per minute (rpms). The circumference along the edge of a carousel, however, is much greater than the circumfer-ence of the hole punched out of its center. That means that the outside of the carousel has to whirl around faster than the inside in order to meet the requisite number of rpms.

So now imagine Moe trying to make his way from the center of the carousel to its edge. As he walks, Moe will find himself being tugged to one side. The same thing, he will find, also hap-pens if he starts at the edge and tries to make his way to the cen-ter. In the first case, the problem arises because Moe is moving slower than the carousel sideways; in the second case, it's because the carousel is moving slower than Moe.

In the case of the earth, the rate of spin translates to one revo-lution every twenty-four hours. In order to comply, the equator, which measures some 24,000 miles around, must spin at 1,000 miles per hour, whereas the poles, like the pole at the center of a carousel, need not move at all. A person standing on the equator, then, only seems to be stationary; like the ground underfoot, that person is actually moving sideways at 1,000 miles per hour.

The Coriolis force is generally perceptible only over large dis-tances. People don't walk or run fast enough and far enough to perceive its tug. But winds and currents and long-range projectiles are a different matter. If ballistic missiles did not take the Coriolis force into account, for example, they would consistently miss their targets.

If one thinks about it, one finds that the Coriolis force accounts for many mysteries, among them, why cool air that flows from higher latitudes back toward the equator does not follow a simple north-south route but is deflected from the east and to the west. The result: the northeasterly and southeasterly trade winds that swoop across the equator, sweeping water before them. The Coriolis force also accounts for the fact that warm air moving from the tropics to mid-latitudes is similarly deflected, giving rise to the westerly winds that are such a distinctive feature of the planet. These winds include the jet stream, and they are called westerlies because, in contrast to the trades, they move from west to east.

The Coriolis force, it is important to remember, always acts in the same straightforward way. The sense of the deflection it causes in one hemisphere is the mirror image of the deflection it causes in the other. In the Southern Hemisphere, it's the deflection to the left that sets up the westerly winds; in the Northern Hemisphere, it's the deflection to the right. In the case of the trade winds, the Coriolis force has the opposite effect: Air flowing north from the Southern Hemisphere is bent right, whereas air flowing south from the Northern Hemisphere is bent left.

Simply because it is the "Line" in between, the equator creates a special playground for the Coriolis force. On either side of the equator, up to a distance of about 500 miles, the Coriolis force has such a slight effect that it might as well be zero. But as winds and currents attempt to spread out north and south of this privileged zone, the sideways tug they experience increases symmetrically on both sides. Water pushed from east to west by the trade winds, for example, is diverted to the right in the Northern Hemisphere and to the left in the Southern Hemisphere. It is, in a word, turned away from the equator as though it were a Janus-faced version of a coastline.

Even more important, insofar as the physics that drives El Niño is concerned, is the effect the Coriolis force has when the trade winds slacken and water starts to move in the contrary direction, from the west to the east. As this east-trending flow attempts to spread out—that is, adopt a slightly more poleward direction—it is constantly steered back toward the equator, channeled along the Line by the Coriolis force in the same way that radio waves are channeled along a coaxial cable. In the language of physics, the equator acts as a waveguide.

THE IDIOSYNCRATIC CHARACTER of the equator began to seem relevant in the late 1970s, after Klaus Wyrtki questioned a small but significant piece of the hypothesis outlined by Jacob Bjerknes. Bjerknes had thought that the warming off the South American coast occurred because the trade winds died down in the east, shutting down the upwelling of cold, nutrient-rich waters from below. Wyrtki, who had spent many years working in Indonesia and Australia, brought a Pacific-wide perspective to the problem. In this particular instance, he was able to demonstrate that Bjerknes was wrong.

To make his point, Wyrtki collected data from existing merchant ships that plied the eastern and western Pacific; he also resuscitated a central Pacific instrumental network that, in the years after World War II, had lapsed into disrepair. The collapse of the trade winds that jump-started El Niño did not originate in the eastern Pacific, Wyrtki established, but much farther to the west. The warming along the South American coast, in other words, demanded a long-distance explanation.

Wyrtki, it turned out, had long been fascinated by the phenomenon of seiches, wind-driven fluctuations in water levels that are sometimes quite dramatic. His interest dated back to the 1950s, when he was a postdoctoral fellow at the University of Kiel in what was then West Germany.

"One winter day," he recounted, "there was very high water in the Baltic, so high that the seaside promenade was flooded. My old professor said, 'Herr Wyrtki, we have to know what is happening,' so we took all the data available, and we found that there

had been a seiche. It was driven by a cold front that had southeast winds ahead of it, moving water from the southern Baltic to
Finland. Then, twelve hours later, as the cold front passed, northeast winds moved all that water back to Kiel."

Seiches are a commonplace occurrence in long, narrow lakes—
Switzerland's Lake Geneva, for example—and for the most part,
they constitute a scientific curiosity, causing nothing more threatening than a rhythmic back-and-forth sloshing. But in large lakes
and shallow seas, high winds associated with passing storms have
been known to create a rogue seiche.

In 1954, for example, a rogue seiche sped across Lake Michigan like an inland tsunami. It was triggered by a line of thunderstorms that moved out over Chicago in the morning, traveling
east at a speed of 55 miles per hour. The winds associated with
these storms plowed up a 5-foot-tall wall of water and pushed it
clear across the lake. When the winds moved on, all the water
they'd piled up in the east came sloshing back toward the west.
Approximately an hour and a half later, on a bright, sunny afternoon, eight fishermen standing on a Chicago pier were swept to
their deaths by a solitary wave; eyewitnesses estimated it to be 10
feet tall.

The equatorial waveguide, as Wyrtki imagined it, was very
much like a long, narrow lake, in this case a lake that was about
10,000 miles long but less than 1,000 miles wide. To the east and
west, the lake was bordered by the hard, reflective rock of continental shelves; to the north and south, the Coriolis force provided
a very real, if less palpable, retaining wall for water flowing east.

In "normal"—that is, non–El Niño—years, Wyrtki observed,
the trade winds pushed water *away* from the eastern Pacific and
piled it up in the west. As a result, the sea's surface measured
more than a foot higher off the coast of New Guinea than it did
off the coast of Peru. But the slope in water level could be maintained only as long as the winds kept on blowing. Once the winds
stopped, the water they'd pushed to the western side of the ocean
had to come sloshing back, rather like a very big seiche. In this

case, the "seiche" caused a slow-moving, nearly imperceptible surge in sea level that propagated from west to east.

What was most important was the effect this sloshback had on the so-called thermocline, an intermediary layer of water that lies between the icy depths and the sun-warmed surface. As the name implies, the defining characteristic of the thermocline is an exceptionally sharp temperature decrease, so that the water at its bottom is some 20 degrees Fahrenheit cooler than the water at its top. Only 100 feet or so thick, the thermocline resembles a permeable plank whose position is not fixed. Off the coast of Peru, where the trade winds continuously push the upper layers of water away, the thermocline skims very close to the surface, while on the other side of the ocean, off the coast of Indonesia, it plunges to depths of 500 feet or more.

The thermocline is important to the physics of El Niño. Colder and denser than the water at the surface, warmer and lighter than the water below, it has a distinct structure that causes it to respond to disturbance all of a piece. In non–El Niño years, the thermocline is depressed in the west and elevated in the east. This imbalance derives from the massive amount of water the winds have removed from the east and piled up to the west. At the onset of El Niño, as all that water comes sloshing back, the weighting across the ocean shifts. Off the coast of Indonesia, the thermocline tilts up, and off the coast of Peru it plunges by several hundred feet. Then, when the winds pick up, the weighting across the ocean reverts to its non–El Niño state.

The seesawing of the thermocline in the ocean is the partner of the seesawing of pressure in the atmosphere. It explains how wind-driven upwelling can continue off the coast of Peru (as Wyrtki had observed) and yet fail to counter the warming of the sea's surface. When the thermocline tilts strongly down, as happens during El Niños, the swizzle stick wielded by the winds can no longer reach the cool, nutrient-rich water that lies beneath it. As a result, upwelling only serves to stir up the sun-warmed water that lies above.

But what causes the thermocline to tilt? The agents, Wyrtki

went on to suggest in that landmark 1975 paper, might well be bizarre, Brobdingnagian waves that were thought to be thousands of miles long. They were seichelike because they were "equatorially trapped" by the Coriolis force, which kept them tightly focused, always bending them back toward the Line.

The type of waves in question had a proper name. They were known as Kelvin waves (after nineteenth-century British physicist Lord Kelvin), and they exerted a strong influence over tides in shallow coastal zones. Shortly before Wyrtki's paper came out, Australian oceanographer Stuart Godfrey had, in fact, suggested that Kelvin waves might be involved in El Niño. In this context, however, they remained hypothetical constructs until Wyrtki used his tide-gauge network to document the broad undulations in sea level that accompanied their passage.

That network was, in fact, in place in 1982. In the eastern Pacific, the Galápagos Islands afforded one measurement point; in the western Pacific, the island of Truk; and in the central Pacific, the islands of Nauru, Kanton, and Kiritimati. By today's standards, however, the tide gauges were extremely old-fashioned. They recorded sea-level fluctuations as a series of punched holes on rolls of paper that had to be collected, shipped to the University of Hawaii, and read by special machines.

Thus, it was only considerably after the fact that Wyrtki and others were able to stitch together the sequence of events that preceded the warming detected by those fisheries researchers off Paita, Peru. From the tide gauges, it became clear that a big Kelvin wave had swept across the ocean midway through 1982, hitting Kiritimati in early July, the Galápagos Islands in late July and early August, and finally, Paita in late September. Then, after it slammed into the continental margins, it veered right and left, propagating up and down the North and South American coast. In response, sea levels rose, the thermocline plunged, and seasurface temperatures shot up from the tip of Chile all the way to the Gulf of Alaska.

FORESHADOWINGS _____

It was November 1996, one of those blustery autumn days when the sky turns to tarnished silver and leaves collect in coppery drifts. Turning out of his driveway in northern Virginia, Ants Leetmaa drove past his neighborhood's woodsy lots, then swung north on the Washington Beltway. At the Camp Springs, Maryland, exit, he angled his ten-year-old Saab onto Auth Road and headed for the parking lot behind the World Weather Building. A satellite dish poked out from the bushes; a jumble of receivers and antennas jutted up from the roof. The grandiose name aside, the most that could be said of this boxy steel-and-glass structure was that it blended well with its surroundings, which included a Day's Inn, a car lot, and a Red Lobster restaurant.

By now, Ants knew, his longtime colleague Ming Ji would probably be sitting in his small, windowless office, sifting through the latest run of the computer climate model that the two of them had spent so long developing. The question that the model was designed to answer was simple: Would sea-surface temperatures in the eastern basin of the equatorial Pacific warm or cool? That is, would there be an El Niño or not in nine to twelve months' time? As always, Ants was more than mildly curious. Not that any one run of the model could be considered definitive, he knew, but the model would sometimes summon up a surprise.

Ants and Ming had been working on the model for ten years. Their first success came in 1992 when the model predicted that a moderate El Niño was on its way. While Ants and Ming did not particularly believe the prediction—it was, after all, the first real prediction their model had produced—they released it, anyway, distributing it to other scientists via a dial-up telemail service. "We got all kinds of messages in response," Ming delights in remembering. "One said, 'You should feel good about your forecast while you can. When it proves to be wrong, your egos will be shattered.'"

After all, El Niños were supposed to recur on time scales that ranged, on average, from two to seven years, and there had already been an El Niño in 1991–92. It just didn't seem plausible that another would so swiftly follow. But the tropical Pacific was in a funny mood, and in early spring of 1993, that was essentially what happened. "Beginner's luck," Ming had said. A short time later, as if on cue, the model betrayed its imperfections. It failed to anticipate a third warming spike that started in the fall of 1994 and lasted into 1995. The spate of closely spaced warmings baffled many scientists. "It's almost as if we've been in a permanent El Niño," Ants said at one point.

Ants had shown up at the World Weather Building in 1985. Ming joined him about a year later. David Rodenhuis, one of Ming's professors at the University of Maryland, had just become the head of what was then called the Climate Analysis Center, and he informed Ming that Ants could use some help. At the time, Ming was a struggling graduate student, newly arrived from the People's Republic of China. All he owned were casual clothes, and suddenly he became concerned. He'd have to dress up for a job interview, right? "Don't worry," Rodenhuis said. "This guy doesn't even wear socks!" Actually, Ants did wear socks most of the time, but in summer he'd sometimes come to work in sandals. "Look," he'd say to Ming, his eyes twinkling behind light-rimmed glasses. "No socks!"

Both Ants and Ming arrived at the Climate Analysis Center at a fortuitous time. In the late 1970s, just before the El Niño of 1982–83 broke loose, Florida State University meteorologist James O'Brien had managed to collect enough data from ocean-crossing ships to concoct a realistic field of trade-wind fluctuations that went back some eighteen years. Then, in 1981, one of O'Brien's students, Antonio Busalacchi, "blew" those winds over one of the early computer models of the equatorial Pacific. Every time the winds shifted, the model responded by producing telltale changes in the structure of the ocean—the height of sea level, for example, or the depth of the thermocline.

O'Brien and Busalacchi presented their results at various meetings. Ants, who at the time was still in Miami, was one of the scientists who attended those meetings. Mark Cane, then an associate professor of oceanography at M.I.T., was another. Ants and Mark Cane had both gotten their doctorates from M.I.T., and while they had not overlapped as students, they were similar in age and knew each other fairly well. "Guys like Mark," Ants once observed, "come up with new ways of looking at the world; they kind of envision things. After that, it's easy."

In the early 1980s, there were ocean models of El Niño, and there were atmospheric models of El Niño. One set of models sought to describe the effects of wind stress on the ocean; the other, the effects of ocean temperatures on the winds. But what would happen if scientists were to couple these models so that changes in the winds would provoke changes in sea-surface temperatures and changes in sea-surface temperatures would provoke changes in the winds? Would such a model reproduce the "chain reaction" that Jacob Bjerknes had envisioned?

In 1985, the same year Ants Leetmaa set up shop in the World Weather Building, Cane and his close colleague Stephen Zebiak started pushing a coupled atmosphere-ocean model into uncharted territory. First they used the model to "hindcast" three events that had occurred in the past, including the El Niño of 1982–83. The model did surprisingly well, so well that they submitted a paper on their accomplishment to the journal *Science*. The paper was rejected. Chastened by the failure of the buildup hypothesis, the scientists who reviewed it deemed that El Niño was an intrinsically chaotic phenomenon, meaning that it was probably futile to try to predict it—or even hindcast it.

Stubbornly, Cane and Zebiak (both of whom had just moved from M.I.T. to Columbia University's Lamont-Doherty Earth Observatory in New York) kept running their model. In late 1985, they noticed, the model was starting to insinuate that a new El Niño might crop up the following year. In January 1986, the model became even more insistent. At that point, they wrote up a new paper, and this time they submitted it to *Science*'s archrival, the British journal *Nature*. This time they included not just the three hindcasts but the forecast as well.

In the spring of 1986, the week of the *Nature* publication,

Cane, Zebiak, and Sean Dolan, a computer programmer who had been working with them, held a press conference. With memories of the failure to predict the 1982–83 El Niño still fresh in people's minds, the warning about false prophets loomed large.

"What if you're wrong?" Mark Cane was asked.

At which point, he shot back, "What if we're right?"

Over the next several months, the warming Cane and Zebiak had so publicly predicted failed to materialize. Then, toward the end of the year, the El Niño of 1986–87 popped up onstage and made a long, sweeping bow.

From the World Weather Building, Ants Leetmaa looked on with interest. The model Mark and Steve had designed was simple and streamlined. It performed remarkably well, Ants thought, considering that it represented only the topmost layer of ocean and employed a stripped-down set of winds to set things in motion. By contrast, the model he and Ming were assembling was clunky. Its ocean was several thousand feet deep; it had twenty-eight layers, and they'd put into those layers every parameter they could think of: current speeds, salinity, pressure, and temperature. But while cumbersome, the larger model had the advantage of being more realistic, or so Ants and Ming were hoping.

Ants and Ming (initially there were just the two of them) had started out by constructing a model of the equatorial Pacific, generously extending it out to 55 degrees of latitude on each side. In essence, they took prominent features like current speeds, salinity, pressure, and temperature and put them into a box. Except, unlike a wooden crate or a cardboard carton, the box Ants and Ming built was a virtual enclosure, constructed out of commands written in lines of computer code. Likewise encapsulated in computer code were selected approximations of physical laws that governed such fundamental processes as heat transfer and the flow of fluids on a rotating sphere.

Every week or so, Ants and Ming would jump-start their model with a fresh infusion of data gathered by ships, satellites, and stationary buoys; then they would step back and watch conditions in their virtual ocean evolve. Beginning in 1990, they began coupling their ocean model to an even larger atmospheric model developed by the National Weather Service. Inside the model, the ocean and the atmosphere now interacted, though not as fully as they did in the real world.

The computer models used to forecast El Niño were often likened to hi-tech crystal balls, and for good reason. Ants and Ming and other scientists really did use them to conjure up glimpses of the future. At first these images were black and white, but later they took the form of false-color maps that spanned the equatorial Pacific from 140 degrees east to roughly 80 degrees west. The hues, arrayed on an intuitive palette, represented patterns of sea-surface temperatures. Blues and greens translated into colder than normal; yellows, oranges, and reds translated into warm.

The transitions in color were very easy to spot, and so in November 1996, as soon as Ants sat down in front of his computer, he immediately latched onto the bright splotch of warmth that the model showed spreading across the ocean midway through the following year. For a few weeks now, the model had been spinning out similar scenarios, and both Ants and Ming were beginning to wonder if it might not be onto something. They were unsure, though; they did not entirely trust their digitized version of a sorcerer's crystal ball.

CHAPTER **FIVE**

Ah, ye admonitions and warnings! Why stay ye not when ye come! But rather are ye predictions than warnings ye shadows?

HERMAN MELVILLE, *Moby Dick*

THE ONSET

THE FIRST WESTERLY wind bursts exploded out of the Indian Ocean in late December of 1996, then moved into the tropical Pacific, sweeping before them a line of storm clouds nearly a thousand miles long. Dawn turned to dusk as violent squalls, fueled by the warmest waters in the world, roiled the ocean and cleaved the skies with ragged lightning. In direct challenge to the reign of the trades, these winds sent water from the western Pacific rushing toward the east.

"Intraseasonals," Ants Leetmaa muttered to himself; he had seen these evanescent disturbances sally across equatorial waters before. A couple of months later, no doubt, some scattered warm patches would light up off the coast of South America, for these wind bursts, just like the sloshback, generated Kelvin waves that raised sea-surface temperatures by depressing the equatorial thermocline. But while these frissons of heat in the ocean were sometimes as dazzling as the light emitted by Fourth of July sparklers, they usually proved just as ephemeral.

There were exceptions to this rule, however. The 1982–83 El Niño, for example, had been so closely preceded by energetic westerlies that, at the time, many scientists wondered if they were more than casually connected. Ants himself, at a workshop held in Miami in 1983, had taken note of the fact that across the central Pacific "wind information from ship reports indicates that close to the equator . . . anomalous westerlies were observed between December 1981 and March 1982." Another even stronger set of westerlies shot across in summer, he noted, and soon afterward, Wyrtki's network of tide gauges recorded the big surge in sea level that swept across the ocean from west to east.

Now, at long last, scientists had the ability to track these events in real time. Shortly after the 1982–83 El Niño, they had installed a network of instrumented buoys in the equatorial Pacific, stringing them out across a distance of 8,000 miles. The first buoy in the network (which is officially known as the Tropical Atmosphere/Ocean, or TAO, Array) went into the ocean in 1985, the last a decade later. In 1997, the equatorial Pacific boasted a total of sixty-nine of these buoys, each one of them anchored to the sea floor by a 2-ton stack of steel wheels scavenged from old railcars.

Painted a gay red and white, these doughnut-shaped buoys sported whirligig instruments known as anemometers for measuring wind speeds and jaunty antennas for uploading data into orbiting communications satellites. But it was beneath the glossy sheen of the water, along the topmost 1,500 feet of nylon cable, that the most important instruments were strung—rugged heat-sensing devices called thermistors that extended all the way from the surface to the bottom of the thermocline. Beginning in late 1995, the thermistors to the west of the International Date Line picked up the presence of a big blob of warmth that lurked, like a pool of magma, beneath the surface of the tropical Pacific.

From December 1996 on, Ants and others intently studied the temperature maps posted on the Internet by the scientists who ran the TAO Array. They also regularly checked the NASA site that kept near–real-time maps of sea-level changes in the tropical Pacific. These maps were produced from data collected by an altimeter on board the TOPEX/Poseidon satellite, whose orbit swung it directly over the equatorial Pacific. Second by second, the altimeter's radar beam pinged the sea's surface along a series of parallel tracks spaced around 250 miles apart, completing a sweep of the sea's surface every ten days.

In February, another series of squalling westerlies moved through, and in March, yet another. In each case, the westerlies produced a Kelvin-wave surge. In March, TOPEX/Poseidon caught red-handed the big rise in sea level that was spreading across the ocean, passing Kiritimati in the mid-Pacific and continuing all the way to the South American coast. Then, in late April, it happened: Sea-surface temperatures rose off the coast of Ecuador and Peru and began spreading north and south. A little bit later, the TAO Array recorded the movement of the big blob of warmth in the western Pacific as it slowly rolled out of position and spilled across the ocean like lava, one awed researcher remarked, from a slowly and majestically erupting volcano.

The westerlies that preceded the warming of the ocean in April 1997 reopened an old debate between those who argued that El Niño is a self-sustaining phenomenon and others, including Klaus Wrytki, who argued that a strong kick of some kind is needed to trigger it. It was a debate that would not soon be settled. "The sloshing of the ocean never goes quiet," Ants said. "Once you get that warm water out by the Date Line, then you get winds that come out and push it. In a way it's like generating an atmospheric storm. You've got to have some sort of favorable condition, and then the atmosphere and the ocean can work together. The winds are always flopping around, but it's not a random flopping. It occurs in a very systematic way."

Right at that moment, Ants was in the process of taking over as head of the Climate Prediction Center, the CPC, as everyone called it. The CPC was the successor to the CAC, the Climate Analysis Center, which had played such a central role in the 1982–83 El Niño. As far as the general public was concerned, the CPC was an obscure organization, nested inside an obfuscating cluster of acronyms that seemed like words in a secret language.

The CPC, those who worked there would cheerfully explain, was part of NCEP, or the National Centers for Environmental Prediction, which, in turn, was part of NWS, the National Weather Service, which, in turn, was part of NOAA, the National Oceanic and Atmospheric Administration, which, in turn, was part of the Department of Commerce. It made visitors' heads swim.

In May 1997, with a powerful El Niño breaching in tropical waters, the CPC uncharacteristically found itself on center stage. That's because it fell to the scientists who worked there to issue what are known as seasonal forecasts—forecasts that attempt to say whether the upcoming summer, fall, winter, and spring are likely to be significantly drier or wetter, warmer or colder, than the year before. The major upheaval that had begun in the tropical Pacific was bound to make a difference. That much everyone knew. What no one knew was whether the CPC would be able to rise to the challenge.

For quite some time, Ants had been critical of the forecasts issued by the CPC, and he had ideas about how to improve them. Indeed, that was one of the main reasons he stepped forward in 1996 when he learned that the position of the CPC director had quite suddenly opened. "I told him he was crazy," laughed longtime colleague Dick Reynolds. "I said, 'Ants, what do you want to do that for? Why do you want to take on all that extra responsibility with no increase in salary?'" Both of them, after all, worked for the government, and as far as Dick could tell, Ants had nothing to gain.

But Ants appeared to have made up his mind, and by then Dick knew what that meant. The two had met shortly after Ants arrived at the World Weather Building in 1985. Dick, who was also an oceanographer by training, sized the newcomer up. Ants, he decided, was sure to fail. Of course, he didn't fail. Ming arrived a short time later. Then came Arun Kumar, an Indian-born meteorologist with a shy demeanor, a whispery voice, and a three-dimensional grasp of weather patterns that seemed almost preternatural. Soon Dick joined the team as well.

By 1990, Ants had become the leader of a small scientific

skunk works that flourished on the eighth floor of the World
Weather Building like a patch of weeds on the edge of a mani-
cured flower bed. The flower bed was tended by the CPC's statistically oriented forecasters, who relied sensibly, it could be
argued, on the time-tested method of seeing in the past the key to
the future. This worked beautifully as long as the future resembled the past but broke down when it did not. By contrast, Ants
had adopted a dynamical approach to climate prediction, the
same physics-based approach that had revolutionized weather
forecasting.

At first, Ants and his team confined their efforts to predicting
the onset of El Niño and its opposite, La Niña. But as they got
pretty good at that, they started producing experimental forecasts
of how swings in ocean temperatures in the tropical Pacific were
likely to affect U.S. weather patterns. Early in 1995, for example,
they churned out a forecast that hinted at an extended dry spell
in the southwestern United States, a consequence of cool La Niña
conditions that had gripped the equatorial Pacific. The dry spell
turned out to be real; it cost the state of Texas something like $8
billion, "and the guys downstairs missed it," Ants reveled in
pointing out. The guys downstairs were the CPC forecasters.

The weeds, in other words, were starting to intrude upon the
flower bed, and some of the gardeners found the encroachment
unsettling. As they saw it, Ants was prematurely advancing an approach to climate prediction whose reliability had not yet been
demonstrated. Sometimes, they noted, a dynamical model might
give the right answer, but for the wrong reasons, just as it sometimes gave the wrong answer for the right reasons.

Ants, though, had a powerful supporter. His dynamical modeling effort enjoyed the status of a special project championed by
Ron McPherson, then the director of the National Centers for Environmental Prediction. McPherson, who had joined the National
Weather Service in 1959, had witnessed firsthand the fierce battles that had erupted over the relative merits of statistical versus
dynamical weather forecasting, and there was no question where
he stood. Over time, he noted, the dynamical forecasting models

(also known as numerical models) had proved their superiority; statistical methods were now relegated to the background, to be pulled out to resolve the ambiguities that frequently arose.

Climate prediction, McPherson felt, needed to undergo a similar revolution. "A characteristic of the development of modern forecasting," he said, "is the tension between people who are more conservative and people who are more progressive." McPherson clearly regarded Ants as one of the progressives. Ants would often wander into his office, McPherson recalled, complaining that the CPC forecasters refused to take the dynamical modeling effort seriously. It was McPherson who backed Ants for the post of CPC director, bypassing two talented CPC meteorologists who also had applied for the job.

By this point, Ants was no longer, strictly speaking, an oceanographer. Like Ming Ji, who had started out as a meteorologist, Ants had transitioned into a climatologist to whom both the atmosphere and the ocean were important. His experience as a graduate student at M.I.T. had served him well in this regard. In addition to Henry Stommel, he had come into contact with other professors of titanic stature. Also at M.I.T. at that time were Edward Lorenz, the architect of chaos theory, and Jule Charney, one of the pioneers of dynamical weather forecasting.

Like Jacob Bjerknes, his own mentor, Charney had developed an abiding interest in the ocean as well as the atmosphere, and in the late 1950s he, Stommel, Lorenz, and others started a tradition of holding joint seminars in oceanography and atmospheric science that alternated between M.I.T. and the Woods Hole Oceanographic Institution, on Cape Cod.

In a metaphorical sense, the dynamical modelers were all Charney's godchildren. It was Charney, after all, who had solved the problem that had bedeviled Lewis Fry Richardson, the gifted English mathematician who generated the first dynamical weather hindcast in 1921. In so doing, Charney deftly swept away the obstacle that for so long had defeated attempts to realize the vision articulated by Vilhelm Bjerknes, Jack Bjerknes's father. In 1904, the same year Sir Gilbert Walker arrived in India, the elder Bjerknes had written:

> If it is true as every scientist believes, that atmospheric states develop from preceding ones according to physical law, then it is apparent that the necessary and sufficient conditions for the rational solution of forecasting problems are the following:
> 1. A sufficiently accurate knowledge of the state of the atmosphere at the initial time.
> 2. A sufficiently accurate knowledge of the laws according to which one state of the atmosphere develops from another.

Some two decades later, Lewis Fry Richardson published a book entitled *Weather Prediction by Numerical Process*. It soon became a meteorological classic. In it, Richardson described his elaborate efforts to hindcast a storm system that had swept across Europe in late May of 1910. To establish the initial state of the atmosphere, he used data on wind speeds and pressure collected by the Geophysical Institute in Leipzig, Germany. He arrayed the data on a grid that divided the atmosphere into some two-dozen cells, each of which measured approximately 120 miles on a side. He then cranked the system forward in time by solving a series of mathematical equations one cell at a time.

Richardson worked on his hindcast off and on between 1916 and 1919, scribbling out equations while perched atop piles of hay in a French barn. (As a Quaker, he was opposed to bearing arms, so he spent World War I serving as an ambulance driver in France.) His patience must have been incredible; it took him weeks to churn out a hindcast for a single six-hour period. "Perhaps some day in the dim future it will be possible to advance the computations faster than the weather advances," he fantasized, spinning out a vision of a futuristic forecast factory powered by some sixty-four thousand human "computers" who solved equations in assembly-line fashion.

Alas, though Richardson got many aspects of dynamical forecasting right, his hindcast scheme turned out to contain a fatal flaw. Near the center of the region of focus, Richardson's hindcast "predicted" a drop in atmospheric pressure of 145 millibars. That was 50 percent larger than the drop in atmospheric pressure associated with the largest hurricanes. Such a pressure drop, in a word, was so improbable as to verge on impossible.

Where did Richardson go wrong? As Jule Charney established, the physics he tried to incorporate was too complicated; it solved equations for extraneous things, like sound waves. Once Charney streamlined the physics, a serious effort to advance dynamical forecasting could finally get under way. By this time, the world's first general-purpose electronic computer, the fabled ENIAC, had appeared on the scene, and in 1950, Charney and his colleagues started using this more powerful version of Richardson's forecast factory to make more realistic weather predictions.

The dynamical approach to weather forecasting did not become a success overnight. It ultimately prevailed because of parallel improvements in scientific understanding, observational capacity, and computing power. Throughout the 1970s, for example, the models used by the National Weather Service produced a plethora of false winter storm warnings. At one point, a prominent meteorologist even urged his colleagues to "abandon research that uses weather sequences in a computer as bases for deduction about the real atmosphere." Fortunately, the modelers did not heed his advice.

Eventually what emerged at the National Weather Service was a close collaboration between the dynamical models and human forecasters. Any one forecast produced by the models represented only what the atmosphere might do; it took multiple computer runs to sort out the most probable scenarios from those far less likely. Even then, meteorologists would sometimes question the results. Like jet airplanes, the models were designed to cruise on autopilot, but there were always instances when a human being had to take over the controls.

Over the course of several decades, National Weather Service forecasters had built up the experience to know pretty much when they should leave the models alone and when they should step in and tweak the results. Something like that was what Ants Leetmaa envisioned happening with climate prediction. It was the only way to make progress, he believed. "You can't hold all this stuff in your brain," he said, "because the system is just too complicated."

In principle, many of the CPC forecasters agreed; in practice, they placed their faith in their familiar statistical tools. Ants knew it would take a lot of persuasion to turn them from skeptics into converts. As the new head of the CPC, he was faced with an urgent need to come up with a viable prediction strategy. He made a pragmatic decision and threw his support to a new, improved statistical prediction scheme that CPC meteorologist Bob Livezey was pushing. The scheme appealed to Ants because it reflected a more sophisticated understanding of how El Niño impacts the atmosphere. Indeed, it reflected everything Bob and his colleagues had learned about El Niño over the past fifteen years. And, as it turned out, it made a superb partner for the dynamical approach to climate forecasting that Ants had helped pioneer.

TELECONNECTIONS _____

The sustained effort to tease apart El Niño's impacts on worldwide weather patterns began in earnest after the El Niño of 1982–83, which, like a slow-moving freight train, transported boxcars of misery around the world. Afterward, Bob Livezey observed, the intellectual landscape changed. "By God, after it was over, we knew what El Niño was, and we knew it had impacts all over the world, and we knew we had to start getting interested in it for real."

In 1982–83, as in 1997–98, it was often hard to decide which was worse—the relentless droughts or the terrible storms that brought so much damaging wind and rain. Thus, along the coast of California, an unfortunate combination of high tides and high winds created storm surges that smashed oceanfront structures, from fishing piers to expensive beach homes. In Malibu, movie star Bruce Dern was spotted sweeping muddy water out of his house, along with splinters of glass from shattered windows.

Meanwhile, on the other side of the Pacific, Australian farmers endured their worst drought in nearly a century. Wheat yields fell by half, farm incomes by a quarter, and bushfires in the southern portion of the continent claimed the lives of seventy-two people and more than three hundred thousand farm animals. The most terrible of these fires erupted on Ash Wednesday of 1983. In the path of the flames lay numerous small Australian communities, among them Ballangeich, Belgrave, and Branxholme. In all, twenty-three hundred houses burned to the ground on Ash Wednesday, despite the best efforts of more than 550 fire brigades and twenty thousand firefighters.

Around the world, the same striking contrast prevailed, not enough water in one place, too much in another. In the northeast of Brazil, for example, a severe dry spell caused crop failures that affected some 14 million people. Other parts of Brazil, though, were plagued by floods of near biblical proportions, with the result that several hundreds of thousands of people in the states of Paraná, Santa Catarina, and Rio Grande do Sul were rendered homeless. In similar fashion, a number of the islands in the western Pacific suffered a drought so prolonged and severe that they ran out of drinking water, while so much rain fell on the central Pacific island of Kiritimati that it flooded out the eggs laid by nesting seabirds.

One of the striking things about 1982 and 1983 was that newspapers and magazines and television reports tended to treat these disasters as separate occurrences, as though each were an outburst of wacky weather unconnected to all the rest. In truth, even climatologists themselves were not entirely sure if they were dealing with fluke congruencies or not. It took several years of pioneering work to show that each of these flagrant violations of climatological business as usual represented a spillover from the same gigantic phenomenon.

Meteorologist Chester Ropelewski, one of the early trailblazers, recalled that he had initially been quite leery of the idea that El Niño influenced weather on a global scale. It didn't help that the phenomenon had been given a name—"teleconnections," literally, long-distance connections—that, to Ropelewski, smacked of something slightly disreputable, "like telekinesis," he recollected with a laugh. Ropelewski was then employed by the Climate Analysis Center, the CPC's predecessor.

But when Ropelewski and a colleague, Michael Halpert, began a methodical examination of more than a century's worth of precipitation records, they themselves were startled by the persistent patterns that emerged. It was as though they had fine-tuned the dial on a staticky radio and suddenly heard a recognizable tune.

El Niño strongly affected rainfall in the following regions, they found: the central Pacific, the south-central Pacific, Indonesia–New Guinea, Fiji–New Caledonia, Micronesia–West Pacific, Hawaii, northern Australia, eastern Australia, southern Australia–Tasmania, central Australia, India, Minicoy–Sri Lanka, East Equatorial Africa, Southeast Africa, northeastern South America, southeastern South America, Central America–Caribbean, and the United States, particularly the Great Basin region and the southeastern sector centered around the Gulf of Mexico.

As if that weren't enough, Ropelewski and Halpert found suggestive hints of impacts on other regions—northern Africa, southern Europe, the Mediterranean, the Middle East—that seemed to blink on and off from one decade to the next.

The paradox was that El Niño's impacts seemed to be both systematic and at the same time highly idiosyncratic. There were always exceptions to the rules. During the 1982–83 El Niño, for example, the monsoon in India was so weak that the rice harvest was seriously compromised; during the 1997–98 El Niño, India recorded an almost exactly average amount of rain. During the 1982–83 El Niño, East African rainfall did not dramatically increase; during 1997–98, Kenya and Somalia suffered the worst flooding in memory. During both the 1982–83 and 1997–98 El Niños, the Indonesian archipelago withered under searing drought and forest fires that had burned out of control; yet during the El Niño of 1965–66, rainfall in this region turned out to be adequate.

And the farther one got from the tropics, the more pronounced the variability became. Both the 1982–83 and 1997–98 El Niños, for example, brought a wet winter to California and a mild winter to Canada and the northern United States. The 1976–77 El Niño, however, had been associated with a very different set of woes. Among other things, California experienced such a severe drought that strict water rationing was imposed on about 6 million households. Eastern states, on the other hand, were assailed by blizzards and subzero cold. Ice glazed the orange groves in Florida and the oyster beds in Chesapeake Bay. More than 16 feet of snow fell on Buffalo, New York. Snow, in fact, fell up and down the East Coast, from Massachusetts to Florida.

An article published by the *National Geographic* called 1977 "The Year the Weather Went Wild."

> From Philadelphia to Peoria, January was the coldest month ever. (In much of Alaska, it was the warmest.) Across the south, cities recorded temperatures that bordered on the absurd: minus 1 degree in Huntsville, Alabama; 10 degrees in Pensacola, Florida; 6 degrees in Jackson, Mississippi.
>
> Snow, too, set records. On January 31 it rested on part of every state of the contiguous 48 for the first time on record. . . . Never had snow fallen as far south as Miami; never had it covered so much of the Northern Hemisphere. By winter's end, 200 inches—nearly 17 feet—had fallen on Buffalo.

El Niño merited only a glancing mention. It was just part of the picture, the *National Geographic* article suggested, more likely the effect of a broader climatological disturbance than its cause.

Even after 1982–83, the variability of the teleconnection patterns remained a vexatious puzzle. In El Niño years, there seemed to be a tilt in the odds of having a wet winter in California and a mild winter in Ohio, but the tilt—in the view of many atmospheric scientists—did not seem great enough to allow for skillful forecasting. For a very long time, El Niño's extratropical teleconnections seemed so elusive that it was tempting to dismiss them as statistical artifacts as opposed to manifestations of real, if rather complicated, physics.

El Niño, it is now well understood, affects weather patterns around the world because it shoves big, important atmospheric features around. In the tropics, for example, it displaces the so-called Inter-tropical Convergence Zone, or ITCZ, which, as its name implies, is the place where the northeasterly and southeasterly trade winds come together. The ITCZ is not a fixed feature of the atmosphere; it restlessly migrates from north to south and back again, following the sun as summer shifts across the equator from one hemisphere to the other. These seasonal swings are particularly pronounced over the continents, and they are important to the monsoonal weather patterns that dominate the tropics.

The reason is quite simple: The convergence of the two wind belts blocks the horizontal flow of air and forces it to rise, creating a zone of exceptionally vigorous convection. Indeed, the reason why some countries that straddle the equator have not one but two rainy seasons is due to the fact that every year the ITCZ passes overhead twice. El Niño does not reverse the sense of this sun-worshiping migration; the ITCZ does not head north when it would ordinarily head south. What El Niño does is redraw the route the ITCZ takes so that it shortchanges some areas in favor of lingering over others.

Knowing this helped scientists explain El Niño's impacts on tropical countries (Kenya and Brazil, for example) that lie quite far distant from the equatorial Pacific. But it did not answer the question that most concerned U.S. forecasters. And that was, how a disturbance that plays out across tropical waters could reach out and touch not just the tropics but also the extratropical world.

In 1983 the answer became clear as scientists analyzed weather charts that depicted conditions not at the surface of the earth but 6–8 miles up, at the jet-stream level. Over the Pacific Ocean, they spotted twin peaks of high pressure that had not been there before. In meteorological parlance, these grand poufs constituted high-pressure domes, or "anticyclones," meaning that they responded to the Coriolis force by rotating to the right (clockwise) in the Northern Hemisphere and to the left (counterclockwise) in the Southern Hemisphere. By contrast, low-pressure troughs, or "cyclones," circle in the contrary direction.

These newly risen prominences lay to the east of the International Date Line, on either side of the equator, and they represented the high-altitude spillover from the convective cloud towers that had moved away from Indonesia and into the eastern basin of the equatorial Pacific. They arose from an equatorial disturbance, in other words, but they were not themselves equatorial features. One sat very close to Hawaii, the other hovered over Southern Hemisphere waters, and their sudden appearance caused a global realignment of airflow patterns through the so-called troposphere, the part of the atmosphere that provides the theater for weather.

In essence, highs and lows are like mountains and valleys. Around them and between them weave the aerial rivers known as jet streams, along with the storms that form in their turbulent narrows and bends. Ants Leetmaa, in fact, often referred to these fast-flowing aerial rivers as "conveyor belts for storms," and what conveyor belts they are, moving high above the surface of the earth at speeds that sometimes approach 200 miles per hour. In large part, it is by rearranging the topography of the troposphere that El Niño alters the flow of weather at higher latitudes.

In an El Niño year, for example, the wintertime pattern in North America is likely to consist of an intense, swirling low right off the coast of California, a ridging of high pressure over western and central Canada, and yet another low over the southeastern United States. The effect of this lineup is to steer storms farther south than usual. By contrast, under non–El Niño conditions, the low in the Pacific is likely to be positioned farther to the west, with a big ridge of high pressure lurking off the California coast. As a consequence, storms are encouraged to flow north, into Washington and Oregon.

Things are complicated by the fact that El Niño–driven changes in atmospheric features always imprint themselves on top of seasonal patterns. El Niño's impacts on the storms that travel along the jet streams, for example, have their greatest effect in winter, when the north-south contrast in temperature is most pronounced. In the Northern Hemisphere, this means December through March, whereas in the Southern Hemisphere, it's June through September. Parts of Chile, for example, experience El Niño in much the same fashion as California, but during the austral rather than boreal winter.

IN 1997, ALL THIS was known. Even so, it seemed that the chaos intrinsic to the climate system might be so great that seasonal forecasts would never rise to a skill level high enough to become socially useful. The concept of chaos many people find difficult to grasp. It doesn't help, certainly, that its scientific definition flies in the face of the popular connotation of the word. But just as the subatomic particles called quarks do not really possess "strangeness" and "charm," so a chaotic system is *not*, as its name implies, disorderly. It only seems that way.

A chaotic system is one whose parameters are all, in principle, measurable and whose physics is well understood, weather being a prime example. It should, in other words, be possible to say what the weather will be like three months from now or six months from now in any region of the world. But it is not. That's because, as a practical matter, not even the most extensive network of meteorological instruments hooked up to the most powerful computer can ever keep tabs on such variables as wind speeds and air temperatures over a single city, let alone the entire world.

The emblem of chaos became a butterfly after M.I.T. meteorologist Edward Lorenz delivered a famous lecture in 1972 entitled "Predictability: Does the Flap of a Butterfly's Wings in Brazil Set Off a Tornado in Texas?" The butterfly's wings were just a metaphor, of course. What Lorenz meant was that small changes in initial conditions could very quickly balloon into very large effects. Mathematically, it could be shown that these tiny flaps, in the course of several days, were enough to nudge the system into a very different dynamical state.

This is why chaos imposes such strict limits on the predictability of weather. Forecasts with twelve-hour lead times, for example, are usually very good, but even so cannot be relied upon to pick up every single small storm that might rip through an area. Forecasts with five-day lead times now perform quite well in forecasting major storms. Forecasts of general conditions, especially cold waves and heat waves, do okay up to about ten days, but that's about it. At two weeks' lead time, a coin flip is as good as a weather forecast, and sometimes better.

For this reason, most front-line meteorologists have long regarded seasonal forecasts with suspicion. CPC meteorologist Bob Livezey was one of the exceptions. He'd gotten his Ph.D. from Pennsylvania State just about the same time Lorenz delivered his famous lecture, and for years afterward, he recalled, "the conventional wisdom was completely dominated by predictability theory. You know, the butterfly's wings? People would say, 'How the hell can you make a forecast beyond fourteen days? You can't. It's a bunch of bunk.' Well, I didn't think it was bunk."

As Bob knew, the atmosphere is a very fast reactor. Molecules of air organize themselves into storms and then quickly dissipate, retaining no memory of the conditions that brought them together. But the ocean is a far denser fluid and so changes much more slowly. Its thermal capacity is immense. As those living along the North Atlantic seaboard well know, the ocean takes much longer to warm up in spring and much longer to cool down in fall, which helps moderate the heat of summer and the chill of winter.

Moreover, there appeared to be a kind of internal metronome

embedded in the El Niño system. According to something called
delayed oscillator theory, the sloshback generated not one but
two types of subsurface waves: Kelvin waves and so-called
Rossby waves, which were named after twentieth-century meteorologist Carl-Gustaf Arvid Rossby. The essential difference was
that these waves traveled across the ocean in opposite directions.
As a result, the east-going waves, the Kelvin waves, were focused
along the narrow waveguide defined by the Coriolis force,
whereas the west-going Rossby waves were allowed to spread
out, like ripples in a pond.

Kelvin waves also moved faster than Rossby waves. They made
a beeline across the midline of the ocean, taking perhaps two
months to cross from one side of the Pacific to the other. The
Rossby waves, by contrast, might take six months or even more.
The really curious thing was this: Upon hitting the solid rock of
the continental shelf on opposite shores, the two types of waves
underwent a striking metamorphosis. Kelvin waves reflected back
as Rossby waves, Rossby waves as Kelvin waves, rebounding
across the ocean like echoes bouncing off a canyon wall.

These echoed waves were important because they were perfectly out-of-phase with the waves that had generated them. Thus,
the first set of Kelvin waves pressed down on the thermocline
in the eastern basin of the tropical Pacific, whereas the second
set—the echoes from the Rossby waves—lifted that weight off,
inviting the thermocline to spring up toward the surface again.
The first set initiated the warming process; the second set brought
it to an end. As Ants Leetmaa liked to put it, every El Niño contains within itself the seeds of its own destruction. And it worked
that way for La Niña, too.

The delay in the oscillator was important: It meant that the
ocean warming was keyed to the length of time it took for these
waves to cross the ocean, around two months for one, six or more
months for the other. These waves were seen as providing the
oceanic component of the cycle with the equivalent of long-term
memory, and as Bob saw it, that ought to give forecasters an edge.
Of course, the cycle that produced El Niño and La Niña would

always be subject to a large degree of chaotic variation; that much was manifest in the irregular rhythms that marked the swings from one to the other. But once the surface of the tropical Pacific warmed or cooled, it could be expected to stay that way not for days or weeks but for a year, maybe more.

There are many climatologically important changes whose time scale dwarfs the window of predictability described by chaos theory, Bob liked to point out. "A volcano goes off in the subtropics that spews a lot of stuff into the stratosphere, and that stuff is going to stay up there for two years. Or maybe half the continent has extremely wet or dry soil for a month or more. Again, that's a change at a longer time scale than fourteen days." As Bob saw it, such changes damped down the chaos inherent in the atmosphere.

To explain why, he liked to call up a mental image of mathematical constructs known as strange attractors. Think of the atmosphere as being organized not around highs and lows but around larger, more intangible objects. Rivers of air, it can be readily imagined, swoop around these objects in much the same way as comets trace highly eccentric trajectories around the sun, brushing uncomfortably close to the earth sometimes.

During a normal year, the large-scale flows in the atmosphere tend to switch from one attractor to another the way a comet does as it swoops through the solar system, its orbit distorted by the gravitational fields of Jupiter, Saturn, and Mars. But during an El Niño year the pattern was more stable. What Bob thought happened then was that these large-scale flows became locked in place around a particular attractor. In fact, he thought El Niño might somehow conjure up a new attractor—a new Jupiter-size planet, in other words—that temporarily grabbed control of the system.

Whatever the mechanism, Bob thought it reasonable to expect that atmospheric patterns, once set up, would exhibit unusual persistence. Seasonal predictability, as he saw it, was not unchanging. It increased during El Niños, most dramatically during strong El Niños. It increased to a lesser extent during La Niñas, and it decreased in the in-between years, the years that people paradoxically insisted on considering "normal."

Still, as Bob had to acknowledge, even the strongest El Niño only damps down the intrinsic chaos of weather; it does not, cannot, eliminate it entirely, which was why climatologists were fond of explaining their forecasts by invoking the image of a game of craps played with a pair of loaded dice. Over a series of throws, those at the betting table will discover that snake eyes are coming up much more often than they had expected. Instead of popping up once every thirty-six throws, snake eyes might pop up one time in ten.

THE WARNING _____

In October 1997, Ants had not completely moved into his new corner office. Cardboard boxes, as yet unpacked, lined the walls, and the surface of a small conference table lay concealed beneath an elbow-deep pile of charts and weather maps downloaded from National Weather Service files. Riffling through this pile, Ants pulled out one of the new forecast maps the CPC had started to post in August. This particular forecast map was for the upcoming winter in the central United States. It was not, Ants emphasized, a weather forecast. It did not aim to forecast rain or sunshine tomorrow or even next week, but rather a winter's worth of rain and sun, warmth and cold.

"What we've done," Ants explained, "is take eleven of the strongest El Niños over the last century and ask, 'How does the rainfall rank?' The greens are wet; the real dark greens—Oklahoma, Kansas, Texas—are very wet; they rank in the twenty wettest years ever. The reds are dry.

"Look at Ohio right here. This is Fremont, Ohio, where I grew up; you can see that the forecast is ranking the upcoming winter there as among the driest.

"These patterns are strongly seasonal and strongly regional.

Look at Mississippi. The northern part is probably going to be dry; the southern part, wet. Far-eastern Maryland could be wet and the western side of the state dry. But in a given event, this could wobble, so while we're saying dry, there's always a potential for wet."

Then Ants pulled out the winter-forecast map for California. The color of the map was extremely dark green. In the past, during strong El Niños, the northern coast of California had chalked up seven years of exceptional precipitation, one dry year, and two years of normal rainfall for the period that began in November and ended in April. The southern coast had a similar track record, with seven extremely wet years, one normal year, and two dry years. The chances that California would be dry during the winter of 1997–98, in other words, were small but not totally negligible.

Ants had just a trace of an accent, a kind of lilting singsong that made one wonder if he had perhaps grown up in Canada. In reality, he'd been born in the small Baltic nation of Estonia, and in 1944, when he was not yet two years old, he and his family had made a harrowing escape from the capital city of Tallinn just as the Russian army launched the invasion that would place Estonia under Soviet control.

On a cloudy afternoon in late September, the Leetmaas had pushed their way through a panicky mob at the harbor to board a decrepit military ship that was part of a convoy organized by the retreating Germans. That night, from the black waters of the Gulf of Finland, those on deck watched as Russian bombers swung low over the city, dropping flares that illumined the sky with phosphorescent streamers. Soon afterward they heard the sound of thunderous explosions.

For three long years, Ants, along with his mother and father and his older brother and sister, drifted across war-ravaged Europe like jetsam and flotsam tossed up by a storm. They landed first in German-occupied Poland, then made their way to Czechoslovakia, and from there to a refugee camp in Augsburg, Germany. In 1947, the Leetmaas boarded a ship bound for New

York. A family photograph shows all three children standing on deck, wide-eyed in winter coats, wearing nametags like pieces of luggage. Ants was five; his sister, Maia, was eight; his brother, Jüri, was twelve.

More than most, Ants knew how violent forces could upend people's lives. The idea that timely warnings might help avert disasters made the CPC's forecasting effort extremely important in his eyes.

"I HAD BEEN going nuts," Bob Livezey cheerfully explained. "We had been dooming ourselves to mediocrity—I'm going to say it in those strong terms—and I was extremely unhappy." As he said this, he broke into a wide grin. Bob is an athletic-looking man with silvery hair, and he was dressed casually, as scientists generally are, in khakis and a blue-and-white-striped shirt without a jacket or tie. Among the forecasters, Bob was Ants's most ardent supporter. To him, McPherson's decision to place Ants at the helm of the CPC seemed inspired.

"Ants," he said, "was a breath of fresh air. He was the perfect catalyst to bust everything wide open." Ants and Bob shared similar views on the nature of science. As Bob put it, "Everything is a moving target, and you've got to have your brain wide open constantly. Almost always when you're thinking in conventional terms, something comes along to knock your legs out from you. You have to be ready to say, 'Well, I was wrong about that,' and forget it and move on. We hadn't had enough of that around here."

Bob and Ants didn't agree about everything, of course. In fact, they argued a lot. Philosophically, for instance, Bob cheerfully agreed with Ants that the future of climate prediction lay in the dynamical models. At the same time, he remained deeply convinced that the power of statistical prediction schemes had not yet been exhausted. That was not surprising, for Bob was a master statistician. He could take a jumble of numbers, sort them into bins, and extract patterns from them in much the same way as a photographer coaxes striking images out of darkroom emulsions.

In 1995, Bob found himself making frequent trips up the stairwell that connected his office on the sixth floor of the World Weather Building to the eighth floor, where Ants and his team were then ensconced. There he found a sympathetic audience for his escalating frustration. What Bob was unhappy about, it turned out, was what Ants was unhappy about—the CPC's failure to use to predictive advantage the physical insights that scientists had amassed since the great El Niño of 1982–83. As a result, both Ants and Bob agreed, the seasonal forecasts CPC had been issuing for most of the 1990s had such marginal skill that they were, in fact, little better than coin flips.

In the course of one such visit, Bob learned that Ants and company had just finished running a very interesting *What if?* experiment. "What they had done," Bob recalled much later, "was drive an atmospheric climate model with a sea-surface temperature pattern that represented a perpetually strong El Niño, and the results really blew me away. It was clear that there ought to be an El Niño signal in the United States in every season, which was completely against conventional wisdom. Conventional wisdom was that there was a winter signal and that it wasn't a very strong signal at that."

This started Bob thinking. It was true that all El Niños were slightly different, just as all thunderstorms are different and all blizzards are different. For one thing, each El Niño started and ended at a slightly different time of the year. For another, each El Niño caused a slightly different pattern of ocean warming. Sometimes the warming began in the east; sometimes it began in the central Pacific. Sometimes sea-surface temperatures stayed nearly normal during one season, only to turn extremely warm in another. But was there a common denominator? A number of scientists were starting to think that there was.

To Bob, the experiment that Ants and company had conducted was highly suggestive. In the virtual world of the model, sea-surface temperatures had not fluctuated up and down as they do

in nature but had stayed uniformly high. In particular, temperatures had stayed high in the east-central portion of the equatorial
Pacific, an area that appeared to be strongly teleconnected to U.S.
weather patterns. Bob liked to refer to it as the erogenous zone.
Perhaps, Bob mused, El Niños didn't really start to matter to the
United States until sea-surface temperatures in the erogenous
zone passed some critical threshold.

It wasn't just a witticism. The erogenous zone was, in fact, a
highly sensitive region, meteorologically speaking. Indeed, it
seemed primed to explode in orgasmic frenzy. During much of the
year, sea-surface temperatures in the area hovered just beneath the
28.5 degrees centigrade, or 83 degrees Fahrenheit mark. That was
the temperature at which the atmosphere really started to pay attention to what the ocean was doing; that was when convective
activity really picked up. It wasn't an absolute threshold but a
relative one. In the erogenous zone, a rise of just 1 or 2 degrees
centigrade was sufficient to create the atmospheric uplift required
to pull the center of intense convection away from the western
Pacific.

It seemed like a good hypothesis. So Bob devised what he
thought was a clever way to test it. First, he decided, he would
sort through prior El Niños month by month, focusing on the
critical region to the east of the International Date Line. Any
month when sea-surface temperatures in this region dipped well
below normal he would toss out. Then he would try to match the
remainder to geographic patterns of temperature and precipitation. The problem was, the leader of his section at the CPC was
not at all eager to pursue the project; in fact, said Bob, "he was
against it." Ants, by contrast, was only too delighted to help.

The result of this impromptu collaboration was an article published in the *Journal of Climate* in August 1997. Its coauthors included Bob Livezey plus Ants and other members of the
dynamical modeling team. What the article argued was that the
United States had a richer, more reliable response to strong El
Niños than had been previously recognized. For Bob, the joint
publication represented a triumphal moment. Even several years

later, as he explained what had happened, he enthusiastically rose on his toes and swept his arms in a wide arc. Statistically speaking, what the CPC had been doing was nonsensical, he exclaimed. It was the equivalent of dumping bags of different-size tomatoes into a blender and mixing them into a paste. No wonder a clearer pattern had not emerged.

In the summer of 1997, with Ants's backing, Bob launched a campaign aimed at convincing his colleagues to adopt a new forecasting scheme, one that concentrated on sorting the tomatoes rather than blending them together. Given the strength of this particular El Niño, he argued, the best strategy would be to throw away all the little cherry tomatoes and select out only the big beefsteak variety, in this case, a subset of El Niños dating back to the late 1800s.

For the period January through March, for example, the following El Niños appeared on Bob's list: 1915, 1919, 1941, 1958, 1966, 1969, 1973, 1983, 1987, and 1992. Gone was the eccentric El Niño of 1976–77. Why? The erogenous zone had not warmed appreciably during that event. "The 1976–77 El Niño is unlike any that we've been able to see in our hundred years of records," Bob puzzled. "We still don't know what the hell was going on."

On the sixth floor of the World Weather Building, Bob and his colleagues argued over the merits of changing their new prediction scheme. "We fought," Bob said, his face wreathed in smiles. Rather quickly, though, the CPC forecasters arrived at a consensus, and truth be told, probably would have in the absence of lobbying by Bob. Sea-surface temperatures had risen so high, so fast, that sheer intuition told veteran forecasters that they were dealing with an extraordinary event, that it made no sense to treat it otherwise.

ANTS WAS IMPRESSED with the CPC forecasters' efforts. He was also impressed by the dynamical forecast turned out by the members of his own former modeling group. That forecast was, if anything, even more insistent that California—all of California, including northern California—was headed for rain. On that basis, Ants started to speak out. "We anticipate that California will be pretty wet," he told *USA TODAY* in August. "It will be very much like 1982–83. It's going to rain a lot. Some of the rain is going to be heavy. There will probably be some flooding; there will probably be mudslides."

Ants said much the same thing in October when he spoke to scientists at the U.S. Geological Survey (U.S.G.S.) in Menlo Park, California. The lecture hall was too small to accommodate everyone who showed up, so he gave the same talk twice. Afterward, U.S.G.S. hydrologists started pulling old landslide maps out of drawers. The maps were so covered with red dots that it seemed as though the entire state had broken out with a bad case of measles.

Not a few veteran forecasters thought that Ants had been incautious. One urged him to qualify his public utterances, invoking the image of driving a nail into one's own coffin. "Never drive the nail in completely," he counseled. "Always leave yourself a little room to pull it out." Indeed, Ants was still getting the hang of being in the public eye. A few months earlier, in response to a reporter's question, he had suggested that people planning winter vacations in California and Florida might want to consider postponing them. Leaders of the tourism industry in both states and their political allies were quick to voice their displeasure.

As October turned into November and California continued to bask under sunny skies, Ants himself began to wonder if he should have spoken a bit more circumspectly. "Failure was not an option for this forecast," he reflected later on. "I really had hammered in the nail." This was the time that skeptics frolicked, and newspaper columnists and television commentators joked about El Nonsense and El No Show. "I laugh in El Niño's face," sneered a columnist for the *San Francisco Chronicle*. "It's more like an annoying chuckle, but I put my hands on my hips and throw back my head so El Niño gets the point."

The skeptics drew credibility from the fact that El Niño's teleconnections to California had for so long been a subject of controversy. "When the historical record is taken into account the correlations [to] California rainfall are not very impressive," two highly respected climatologists had written in 1983, and fifteen years later, among many meteorologists, that view still prevailed.

At the CPC, however, there was no question about the power of this El Niño. The Indian monsoon might not have failed, but fires were breaking out in Sumatra and Borneo, sending plumes of choking fumes into Singapore and Kuala Lumpur. Villagers were fleeing the terrible drought that had engulfed the highlands of Papua New Guinea. And Chile, South America's equivalent of California, had been pummeled by powerful winter storm systems, while an eerie quiet had settled over the Atlantic Ocean, where wind patterns associated with El Niño ripped the tops off developing hurricanes.

And all this, veteran El Niño watchers well knew, was only the beginning.

Besides, this time both the statistical and dynamical forecasts had come into such alignment that, for the first time really, Ants had confidence that the increase in precipitation he and his colleagues had forecast would, in the end, prove out. A 30 percent increase in precipitation might not sound like much, he observed; the problem was, that increase would be packaged in the form of exceptionally violent storms. Raindrops were not likely to fall gently from heaven; they would be borne by savage winds that would drive them into the ground like spikes.

Throughout the fall of 1997, then, Ants did not waver; he continued to speak out, even though he understood that there would be a price to pay should the forecasts be perceived as wrong. That price would be measured in the loss of public confidence and, possibly, public funding as well, and the CPC, Ants knew, could afford to lose neither.

For about three years, along with the rest of the Weather Service, the CPC had endured a regime of federal budget cutting so draconian that for a time its scientists lacked travel money to attend conferences in their field. Collectively, they were demoralized, and to compound their sense of insult, their plight seemed to garner very little sympathy. Meteorologists at the Weather Service, for the most part, considered the CPC unessential. Meteorologists at universities deemed it unworthy. The scientists working at the CPC had a lot riding on the forecasts they had just made.

In November 1997, Ants appeared with NBC *Today* show weatherman Al Roker on CNN's *Larry King Live.*

"Do you worry, Ants, that, you know, we've talked about this so much and said that 'Hey, this thing is coming,' and then if we don't get these horrible storms . . . you know how people are, 'You blew it! You said it was going to happen, and now it didn't happen.' And they're mad they didn't get the bad storms?" Roker asked.

"You're right. I lose a lot of sleep over that. I'm kind of waiting for this pattern to set up. I have faith in physics," Ants said.

That faith would be tested for two months more. Then, finally, in January, the anticipated pattern set up, and the storms started rolling in, one after another, just like the bombers that had attacked the Estonian capital of Tallinn on that night so long ago.

CHAPTER **SIX**

The huge Leviathians to attend their prey,
So close behind some promontory lie
And give no chace, but swallow in the fry
Which through Gaping Jaws Mistake the Way
                JOHN DRYDEN, *Annus Mirabilis*

STORMY WEATHER _____

Had he been flying above the clouds rather than through them, Marty Ralph would have seen the sun rise above a swirling stormscape, bathing airy battlements in pale rose light. As it was, he only dimly sensed the transition from night to day, for dawn, when it finally arrived, came cloaked in a miasma of gray. Several hundred miles to the east curved the craggy coastline of California; more than 2,500 miles to the west lay the island chain of Hawaii. Directly below was only ocean, and it was wild with waves.

Somewhere down there someone might be in trouble, Marty thought, and at that moment, the vintage turboprop that carried him started to pitch, just like the ship that, in his mind's eye, he had imagined was being tossed about by frenzied seas. Granted, Marty had come to California in the winter of 1998 expecting that El Niño might well put on a spectacular show, but even so, the vast cyclonic systems that he and his colleagues flew into that first week of February were more powerful than anything they'd dared dream of.

No winter storm that slams into the West Coast of the United States materializes out of nowhere, of course, and in that sense these storms were no different from any others. They took shape out in the central part of the North Pacific as a mass of frigid air swooping down from the Aleutians collided with warm, wet air spiraling up from the south. Like a pair of sumo wrestlers, the opposing air masses warily began to circle each other, eventually forming an expanse of cloudiness that encompassed an area twice as large as California itself.

At the leading edge of these storms, a thin cirrus veil imparted a foggy cast to the atmosphere and surrounded the sun with a pellucid aurora. Behind the cirrus lurked a zone of tempestuous winds that rotated in a counterclockwise direction around the intense zone of low pressure at the storm system's core. It generally took several days for storms like these to make landfall, and so, well in advance of their flights, Marty and the others had anticipated that disturbances of tremendous power would be heading their way.

Marty had arrived in California around the middle of January 1998, when a beguiling series of sun-drenched days made it hard even for him to believe that the sky would soon fill with anything more sinister than a regatta of cotton-puff clouds. Up and down the coast, not far from the rambling house that Marty and two colleagues had rented in Carmel, children raced along sandy beaches. Starfish with nubbly arms crawled along the bottoms of rocky tide pools. Seagulls swooped, filling the air with raucous music. All seemed as it should be save for the waves that toward the end of the month started to roll in high, very high, signifying that a great commotion had commenced out to sea.

The flight on February 2 marked Marty's eighth foray on the P-3 Orion, and each time he found he was becoming more comfortable with both the plane and its crew. Initially developed for the navy to scout for submarines, the P-3 excelled at flying low and slow over the ocean, in all kinds of weather, and since 1976, the National Oceanic and Atmospheric Administration had kept two on hand, stationing them for much of the year at MacDill Air Force Base, near Tampa, Florida. The main job of these stalwart planes was to fly scientists into hurricanes, but when there were no hurricanes to chase after, the P-3s were placed in the service of other branches of meteorological research.

The P-3 to which Marty and his team had been entrusted boasted a wingspan of 99 feet, nearly equal to its 116-foot length. That made it a skillful glider capable of remaining airborne with the engines cut off for considerable amounts of time. It also boasted an impressive range. At lower altitudes, where the air is thicker and offers more resistance, the P-3 was said to be able to fly more than 2,500 miles before refueling; it could cover a distance nearly twice that far in the ethereal reaches of the upper troposphere, which during wintertime at the latitude of northern California tops off at roughly the same height as Mount Everest.

In keeping with its status as a flying laboratory, the P-3 was equipped with computer work stations connected by cables to instruments mounted on the exterior. There were three radar units: a conventional unit on the black nose cone, another on the white-painted belly, plus a boxy Doppler on the tail. There were gauges on the fuselage and wings that took continuous readings of humidity, pressure, and temperature, and jutting out from the nose cone was a beaky spike that doubled as an air sampler and in-flight lightning rod, so that in profile the plane resembled an outrageously outsized hummingbird.

Marty and his colleagues boarded the P-3 around two o'clock in the morning on February 2. They heard the turbines power on with a high-pitched whine. They felt the fuselage shake as the propellers started spinning. They sensed the plane start down the runway, rising into the air with a touch of tentativeness, or so it seemed. From the small airport in Monterey, California, the plane followed a course that was nearly due west. It angled up and over the giant kelp forests of Monterey Bay. It gained still more altitude over the bay's submarine canyon, where the seafloor drops off to abyssal depths just 60 miles offshore. Then the P-3 headed out over the open ocean, the domain of great white sharks, giant sunfish, and schools of sleek, keen-eyed tuna.

As chief scientist for the flight, Marty split his time between the cockpit and the cabin. Whenever the plane went into a steep bank, he felt a sudden surge of centrifugal force that made walking strangely laborious, like trying to move through thick molasses. What he was trying to locate was the storm system's low-level jet, a fast-moving river of warm, wet air that skims just above the surface of the foam-flecked ocean. To find it, he and his colleagues methodically pushed a series of instrumented probes through a hatch on the floor of the P-3. Known as dropsondes, these probes radioed back a steady stream of meteorological data during the time it took for them to parachute down to the water.

Around dawn, as the plane cruised along the southeastern boundary of the sloped-back cold front, an audible whoosh announced the fall of the tenth dropsonde. On the basis of the data it relayed back, Marty directed the pilot to begin a gradual descent. "This is it!" others on board heard him exclaim as the P-3 bumpily circled down, down, down, tracing wide arcs as if descending an invisible staircase. The plane was right on top of its vaporous quarry.

By midlife, many people fall out of love with their chosen profession, but Marty was not among them. He continued to be as mesmerized by weather now as he had been at the age of twelve, after his family moved from Michigan to Arizona. In suburban Detroit, houses and trees obscured the sky, but on the outskirts of Tucson, the horizon was so uncluttered that storms could be seen coming from a hundred miles off. Sometimes rain would fall from the sky, then evaporate before touching the ground. At other times, it would pour in torrents, triggering flash floods that gushed through narrow arroyos. Rain, Marty was astonished to discover, did not just feel wet; it smelled wet, and so whenever he saw clouds approaching, he began to wait expectantly. At some point, he stopped watching clouds and began studying them. Like an entomologist trying to catalog species of butterflies, he began to associate names with the structures he saw: nimbostratus, cumulonimbus, stratocumulus, cirrostratus.

Over the years Marty had developed an intuitive feel not just for clouds but also for the seemingly vacuous layers of air that lay above, below, and between them. It was in these as yet cloudless spaces that Marty sensed the danger of this storm. The P-3 was now flying through the so-called boundary layer, a section of the atmosphere that extends from an altitude of a few thousand feet right to the surface of the storm-stirred ocean. The boundary layer wraps the planet in something akin to skin; it is within this layer that the atmosphere "feels" whatever lies beneath it, whether a rocky continent or a tempestuous ocean. Through the boundary layer rushed the low-level jet, and as soon as the P-3 entered the jet stream, Marty could see air so laden with moisture that it looked murky, almost steamy. Once that gravid air slammed into the coastal mountains, Marty knew, it would give birth to buckets of rain.

It was on the way back to Monterey, as the P-3 rounded a broad bend in the coastline, that the first squall line hit. The plane touched down amid wind gusts that topped 50 miles per hour. To Marty it seemed as if the storm were just starting to wind up, like a pitcher about to unleash a fastball. He sucked in his breath. This storm system was going to put on one hellacious show. With the next flight scheduled to go just fifteen hours later, he decided to drive back to Carmel and get some sleep. By the time he tumbled into bed, it was midafternoon. The sky was morphing into an inky whirlpool. The wind was starting to moan. Just before he dozed off, the power went out.

Around nine o'clock that night, Marty awoke to glowering darkness. He pulled on jeans and a turtleneck, ran a comb through his straight blond hair, grabbed a flashlight, and made his way to the kitchen. At first he thought he'd fix something simple to eat. There were at least a dozen cans of soup in the cupboard, he knew, plenty of pots and pans, even a box of matches to light the gas stove. But where was the can opener? After fumbling through every drawer, Marty gave up. He grabbed his jacket, sprinted through pelting rain to the rental car he'd parked on the street, and headed off to look for a restaurant.

It proved to be a memorable drive. Hurricane-force winds had sheared heavy branches off sculpturesque cypress trees. Power lines sparked as they draped across roads. Streetlights were out. Stores were dark. Finally, Marty spotted a brightly lit island that housed a Bakers Square restaurant. Joining a contingent of exhausted policemen and bleary-eyed electrical crews, Marty slid into a booth and ordered soup, salad, and a big piece of pie. Afterward he got back into his car and continued on to the National Weather Service command center in Monterey, where forecasters were fielding calls from emergency managers across the whole of the San Francisco Bay area.

As Marty and the other scientists drew up their flight plans, stream gauges throughout the area were hitting flood stage, and alarms were going off in the command center, one after another. Suddenly, the lights flickered, and conversation stopped as an emergency generator strained to keep the power flowing. "Everyone was just holding their breath," Marty remembered, "hoping that the generator wouldn't fail."

Around three-thirty in the morning on February 3, the P-3 took off again. This time it headed southwest, then east, threading a course off the coast of Santa Barbara, where winds were gusting up to 70 miles per hour. To one side lay the sheer cliffs of the Channel Islands, to the other the low mountains of Southern California. All of a sudden, the navigator looked at his radar screen and allowed as how he wasn't absolutely sure if he was looking at rock or very dense clouds, at which point the pilot made a judicious U-turn and headed back out over the ocean.

By then, the Los Angeles airport had shut down, so that for several hours the P-3 flew through these stormy skies alone. Later in the day, on its way back north, the P-3 nearly collided with a looming line of thunderstorms. Their shapes reminded Marty of the towering cloud formations that spawn tornadoes in the South and Midwest. Far below, where rivers emptied into the sea, he spotted long, dark shapes bobbing in the water, trees, he guessed, uprooted by landslides. Somewhere down there, he again thought, someone might be in trouble.

THE FOURTH OF FEBRUARY was notable for its shifting mosaic of sunshine and shadow. Waterfalls spilled over seaside cliffs, rainbows arched over clouds, and a new swarm of giant waves leaped up from the sea, outriders of the next storm bearing down on the coast. National Weather Service forecasters were somber. "Our district is in trouble," began one bulletin issued by the Monterey center, repeating Marty's refrain.

Early on the morning of the fifth, two members of Marty's team headed north on the coastal highway to launch weather balloons from the ground and download data from ground-based wind profilers. It was already raining quite hard when they reached the vicinity of Cazadero, a town that lies slightly to the northwest of the village of Rio Nido. In fact, Cazadero Creek, one of the many tributaries of the Russian River, was already nearing flood stage when Marty and his colleagues took off in the P-3 again.

Ola Persson, a tall Swedish American with a halo of curly hair, was the chief scientist for this flight. He vividly remembers how the pilot brought the P-3 so close to the ocean that it flew through the spray tossed up by breaking whitecaps. From a big bubble window, Persson managed to shoot a photograph of the seething sea below; it looked like a sheet of corrugated metal diagonally streaked with white. The plane's instruments recorded wind speeds near the sea's surface that approached 100 miles per hour.

While flying south on the way back to Monterey, Marty once again spotted long, dark objects floating along the coast. The next day, he walked from the house where he was staying down to a nearby beach and beheld a jumble of redwood and Douglas fir that looked like giant pickup sticks. The trees, he knew, had fallen somewhere in the mountains and had been swept by swollen, surging rivers down to the ocean. "They were stacked as deep as this room," he told his wife, Liz. There was awe in his voice. A short time later, he picked up a newspaper and read about the Rio Nido landslide.

FOR MARTY AND his colleagues, the next two months dissolved into a blur of activity. They didn't return from California until the end of March, by which time they had made about two dozen flights, four times as many as they originally anticipated. In a La Niña year, these big winter-storm systems would have been far more likely to follow a northerly trajectory, causing flooding and landslides in Oregon and Washington. Thanks to the 1997–98 El Niño, however, most of these systems made landfall farther south, turning California into a laboratory for storms.

The storms that caused so much misery in North America during the winter of 1997–98 were all extratropical cyclones. They were, in a word, the same storms whose power had mesmerized Jack Bjerknes before he turned his attention to El Niño. "Like pearls on a string"—that's how Bjerknes referred to them—these powerful disturbances frequently arose one after another along the looping battlefronts defined by the jet streams, the subtropical jet stream as well as the subpolar jet stream, which is known as the polar front jet. Each of these "pearls" was enormous, spreading across 1,500 or more miles, so that a blizzard might engulf New England at the same time that tornadic thunderstorms were blasting the Southeast.

Until the eighteenth century, it is worth noting, scientists had little inkling that meteorological disturbances of such immense scale existed. For the most part, it was assumed that storms had more or less local origins. In 1743, however, the indefatigably ingenious Benjamin Franklin took a giant step toward shattering this long-standing assumption. On the night of October 21, he had been planning to observe a lunar eclipse when a storm obscured the skies over Philadelphia. To his surprise, he later learned that Boston had been hit by a similar storm, but on the following day, so that the eclipse had been clearly visible.

Franklin very quickly deduced that Philadelphia and Boston

had been struck by the same storm at slightly different times. Piecing together accounts of the storm from travelers and newspapers, he determined that the storm had probably formed in the vicinity of Georgia and the Carolinas, then moved up the East Coast. But there was a puzzle. The winds in the storm had blown from the northeast, which should have made the storm hit Boston before it hit Philadelphia.

Franklin could only grope for an explanation. Possibly the storm had been fueled by the rise of warm air "in or about the Gulph of Mexico," he conjectured, so that cold, dense air from the northeast could be visualized as rushing toward a zone of low pressure positioned far to the southwest. Today, of course, we know that Franklin's explanation missed the mark. It is the counterclockwise rotation of warm- and cold-air masses around a low-pressure core that generates the wind patterns typical of this class of extratropical cyclones, which are popularly known as nor'easters.

Extratropical cyclones are much larger than hurricanes and structurally much more complex. Hurricanes, after all, are just rotating cylinders of storms that wrap around a clear central eye as if it were a chimney. They are marine tempests that draw sustenance from warm, moist oceanic air that rises and condenses, thereby releasing bursts of heat. Anything that distorts the vertical structure of the storms that form the walls of the chimney or disrupts the upward flow of air will cause a hurricane to lose energy. This routinely happens whenever a hurricane encounters a patch of strong winds, colder water, or modestly rough terrain.

Extratropical cyclones, by contrast, can cross an entire continent before petering out. That's because they derive energy from the combat between warm- and cold-air masses, and they flourish in the fall, winter, and spring, when the contrast between these air masses is greatest. Sometimes fast-moving warm air will attempt to slide over the top of a more sluggish cold-air mass. At other times, cold air will shove its way beneath a slower-moving warm-air mass. Either maneuver is capable of triggering a big outburst of rain, snow, sleet, or hail as long as there is a core of intense low pressure at the developing cyclone's center.

Extratropical cyclones invariably break out along the same wavy path that the jet streams follow as they circle the world. The reason: At high altitudes, the jet streams mark the division between cold-air masses (which subside to form low-pressure troughs) and warm-air masses (which balloon out to create high-pressure ridges). Also important to the development of these storms are so-called jet streaks, bands of exceptionally high winds within the jet streams. The jet streaks assist the development of extratropical cyclones by scouring even more air out of the central core of low pressure, and as air rushes in to fill the void, the winds pick up, and the storm intensifies.

The jet streams (and the jet streaks within them) do not follow straight-line trajectories; like rivers, they can split into branches, or equally well, they can merge. The confluence of two jet streams can produce storms of exceptional power. During the modest 1993 El Niño, for example, a superstorm that derived energy from both the subtropical and polar-front jet streams became the most infamous extratropical cyclone in modern history. That storm had a central core of low pressure that invited comparison with a class 3 hurricane, and it slashed a path of destruction from Mexico and Cuba to the Southeast and New England.

Among other things, the March 1993 superstorm caused more than fifty tornadoes in Florida and whipped up waves mighty enough to sink a 600-foot freighter off the coast of Nova Scotia. It blanketed Atlanta, Georgia, Asheville, North Carolina, Washington, D.C., and Providence, Rhode Island, with wet, heavy snow that was often mixed with sleet and rain. On Long Island a storm surge powered by 90-mile-per-hour winds claimed nearly twenty homes even while the snow was falling. By the time it was all over, the death toll stood at 270.

In the atmosphere, as in all realms governed by physics, anything that is not forbidden is allowed (though some things, granted, are more probable than others). Thus, extratropical cyclones can form along the subtropical jet stream in the mid-Pacific and barrel into California, or they can form along the polar-front jet and broadside the state of Washington. They can form along the polar-front jet on the lee side of the Rockies and blast across the Midwest, or they can originate along the subtropical jet stream in the environs of Georgia or the Carolinas and move up the East Coast, just like the nor'easter Ben Franklin so meticulously described.

What is different about an El Niño year is that extratropical cyclones exhibit a very strong bias. They tend to form along the subtropical jet stream, which, due to the warmth being pumped out of the tropics, tends to be exceptionally energetic.

Usually the subtropical jet stream keeps well to the south, but occasionally it will loop very far north. That, in fact, happened in January 1998 when an extratropical cyclone spun through New England and Quebec in very slow motion. In Canada alone, the freak ice storm that resulted caused at least twenty-five deaths, cut off power to a million homes, brought down 70,000 miles of power and phone lines, felled thirty thousand utility poles, crumpled more than 120 high-voltage transmission towers, and laid waste to millions of trees, including vast groves of stately sugar maples from which maple syrup comes.

More typically, the extratropical cyclones that form in El Niño years slash a path that takes them much farther south. In late February 1998, one such storm, tanked up with warm, moist air from the Gulf of Mexico, spawned a swarm of tornadoes that touched down in central Florida, killing 42 people and injuring 360. In the space of just a few hours, some seven tornadoes swept through four Florida counties, venturing uncomfortably close to major tourist attractions, like Universal Studios and Walt Disney World. Winds in excess of 180 miles per hour lifted up roofs, blew down walls, and twisted the aluminum frames of trailers and campers beyond recognition.

Among the worst was the tornado that loomed up southwest of the city of Kissimmee, Florida, in the wee hours of February 23, when almost everyone was sleeping. First it staged a surprise attack on the Ponderosa RV Park, near the spring training camp of the Houston Astros. In the space of minutes, it plucked people from their beds and hurled them outside, along with their household possessions. One man was knocked in the head by his own TV set; he found his wife on the ground, alive, with a big splinter of wood sticking out of her back. A short time later, the same tornado shattered a small frame house in a semirural neighborhood and sent a toddler sailing into the tangled branches of a fallen oak tree. Half an hour later, a search party found the eighteen-month-old boy still nestled in his mattress and blanket; miraculously, he escaped serious injury.

At that very moment, yet another one of Bjerknes's "pearls on a string" was dumping an extreme amount of rain on the Southern California community of Laguna Beach, unleashing a spate of landslides. A report issued by the California Department of Mines and Geology tallied up the damage:

> On 2/22 Laguna Canyon Road was closed due to mudslides. On 2/23 in Canyon Acres a house partially collapsed under a slide, and part of a bungalow on Victory Walk was crushed. On 2/24 the rainfall forced 550 people from their homes. A wall of mud roared down hillsides, killed one man, injured nine people, and destroyed four homes in the Castle Rock community. One man was killed when a mudslide slammed into his cottage on Victory Walk. On Canyon Acres Drive, three houses were destroyed and five others damaged.

The storms of February 22 and 23 served as a kind of climax for the El Niño winter of 1997–98; none of the storms that formed afterward had quite the same punch. From Florida to California, people affected by the storms started to put their lives back together again. That included the residents of Rio Nido, California, all of whom had fortuitously lived through the landslide touched off by the storm of February 5 and 6.

"No marlin! No marlin!" Phyllis LaCombe was yelling at the very top of her lungs. She was laughing, too. It was well into March 1998, and Phyllis and her husband, Gary, were moving everything that they possibly could out of their ruined home, even the trophy fish that Phyllis not so secretly loathed.

Sunshine, more intense than anyone could recently remember, streamed through the ferny needles of nearby redwood and Douglas fir and sparkled in Phyllis's long brown hair. She was dressed in jeans and a sweatshirt. In her boots she stood about 4 feet 11 inches; in reality, she was only 4 feet 10 inches. In her determination to get the job of moving out done, over, finished with, she seemed extremely imposing somehow. Other than rescuing the dreaded marlin, no one failed to do what she asked.

Phyllis grimaced as she walked through the muck that now covered the white vinyl floor of her kitchen. "What about the fridge?" she wondered. Its door hung open, revealing a row of eggs more than a month old. The oven of her brand-new stove had accumulated a load of mud. "I hadn't even baked in that stove," Phyllis protested.

Soon both the stove and the fridge were trotted outside by a phalanx of friends, who temporarily left them on the raised plywood walkway that had replaced the road. The walkway sported big orange arrows that pointed uphill, to safety—in case another slide let loose, a county official explained—and big orange letters that spelled out the word ESCAPE. Seemingly oblivious to the irony, convicts from a nearby state prison worked to channel the flow of mud away from the walkway and into a sandbag-lined trench.

Over the course of a couple of hours, the LaCombes' house disgorged its considerable contents. Cans of tomatoes, peas, and chicken-noodle soup joined tables, lamps, and chests of drawers on the plywood walkway. A mud-stained oak hutch and a couch with baby blue upholstery sat side by side with a bright white toilet. From time to time, someone would sit on the toilet's closed seat as though it were a lawn chair. There was a surreal quality to the scene, as if the LaCombes and their friends were actors in a Federico Fellini movie.

High in a bedroom closet, safe from the mud, Phyllis found a cardboard box that contained a baby's white piqué baptismal dress. "I've been looking for this," she said, sounding pleased. Gary, in the meantime, had dug out the all-leather Mad Hatter's hat he'd made and worn at a long-ago Halloween party. He perched the hat on top of his head and grinned.

"Hey, Gary," neighbor Pat Kelly called from somewhere in the house. "I saved the book on how to keep your pool sparkling clean."

Everyone giggled. The LaCombes did not have a pool, and nothing in a house being swallowed by mud could be called sparkling clean.

In very slow motion, the slide that started in early February was still continuing, for above Upper Canyon Three hung a huge, unhinged block of earth that with each new rain sloughed off rocks and mud. Worried that an even greater catastrophe might be in the offing, county officials ordered the evacuation of some fifty homes and then expanded the order to include nearly one hundred more. As a spokesman for the California Office of Emergency Services put it, "The problem is that we don't know when the other shoe will fall."

For the LaCombes, the shoe that had already fallen was more than enough. Gary realized that in the chaotic first days after the slide, when he slipped past the guards who had been stationed to block access to Upper Canyon Three in order to retrieve some personal effects. It was nearly dusk, he remembered. Rain was falling; another river of mud was flowing downhill. Nervous about the possibility of a new slide, the guards had all left their posts. All of Upper Canyon Three seemed deserted as Gary made his way across a foot trail that angled across higher ground, then over the logjam and into his house.

This was the first time that Gary saw how his house had spun off its foundations; it was the first time that he understood the power of the slide, and being there all alone made him nervous. He was beginning to haul out some treasures, including some $400 worth of quarters, when he ran into neighbor Doug Spencer. At first, Doug did not recognize Gary and mistook him for an intruder. "Shut up, Doug." Gary laughed. "I own the damn house."

In the time since, mud had completely covered the cars that he and Phyllis had so casually left in the driveway; it was piling up ever higher against the first floor of the house, oozing into the downstairs through cracks large and small. The aptly named mudroom lay under a goo of liquefied earth. The living-room carpet had turned to chocolate-colored sludge; it squished underfoot like a waterlogged sponge.

Even more alarming, the flooring beneath the carpet had begun to heave, and the stairs had acquired such a steep tilt that those brave enough to climb them felt faintly seasick. Plus the sliding-glass doors in the living room no longer opened; the frames had warped, and the glass had bowed. Soon, Gary worried, the panes would implode with lethal force, unleashing a barrage of sharp-edged projectiles.

In just a few weeks, the LaCombes' cozy house had become an alien, even hostile, place. The air inside smelled dank; freckles of mold crept up walls; windows wept with mist. The entranceway was just a jagged slash in the wall of the kitchen; during the move, someone had grabbed a chain saw and unceremoniously widened it. This act of casual vandalism, this breach of everyday etiquette, underscored the outrageousness of what had just happened. Nature was now the owner of this house, and there was no point in pretending otherwise.

The tremendous alteration that had taken place was evident everywhere, inside and outside the house. The backyard, with its gazebo, flowering trees, and spring bulbs, lay buried beneath a mire of mud and wood. Climbing atop the trunk of one tremendous fir, arms spread wide for balance, Gary wonderingly surveyed the scene. His plaid shirt and jeans had turned to khaki; his mustache drooped; he looked as tired as a mud wrestler who'd lost a big match.

ACROSS THE WAY, Hee-Ran Schultz stood in her yard smoking a cigarette. She had always been slender. Now she looked haggard. Her eyes glittered with tears. "I came here thinking no more crying, no more crying," she said. "But this is hard. This is not like a flood. A flood is maybe a couple of days and then the water's down and you can clean up. But the mud just stays there. It can't go anyplace. Maybe it's going to cover up the house."

To Hee-Ran, the slide seemed, above all, unfair. She worked in the production facility of the nearby Korbel winery, and she'd scrimped and saved in order to buy her house, which, at the time, had cost about one hundred thousand dollars. Bit by bit, she'd added new furnishings. Just before the slide, she had bought a pretty oak buffet and two baby blue couches. Now the house was so mud-logged that the doors of the buffet had warped, and the bottom of the couches looked as if coffee had been spilled on them.

Hee-Ran had been terribly proud of her home and of the steep hill behind it that she'd transformed into a wonderland of flowers. "Flowers, oh, God, everywhere there were flowers," she said. "I made layers of different colors, yellow and pink and white, so that everyone passing by could see them. In the springtime, it was so beautiful."

Forty or so feet above the roof of the house, someone had sprayed orange paint across the trunk of a tall redwood tree to mark the exit route should another landslide spew forth; if all the mud at the top of the canyon came down, Hee-Ran shuddered to think, it would probably cover everything below the tree.

Mud had already buried the ten rosebushes she'd just bought and the hydrangeas that grew alongside the house. But higher up on the hill, spring was coming, and primroses and daisies would soon swell with buds. Hee-Ran bent over to pull out a weed. Even then, after all that had happened, it was still her garden.

The day the LaCombes moved out was Hee-Ran's fiftieth birthday. That afternoon, around four-dozen people gathered at the old Rio Nido fire station to throw her a surprise party. Hee-Ran came dressed in mud-streaked jeans and a navy blue sweatshirt, and she wrapped her arms around her body as if she were cold. When everyone sang "Happy Birthday," Hee-Ran cried.

IN THE WEEKS following the slide, the fire station, a cavernous cinder-block building, turned into a kind of community center for the displaced residents of Upper Canyon Three. Even on Sunday afternoons there'd often be a group of people sitting around just talking. Among the regulars were Hee-Ran and her daughter, Mai Lin, their neighbors Jim and Paula Kudroff, and volunteer fireman Pat Kelly.

Pat Kelly lived just down the street from Hee-Ran and the Kudroffs as well as the LaCombes. The chain-link fence the county had erected to cordon off the danger zone started just below his house. It drove him crazy, he said, to see security guards casually strolling around his property at all hours when he himself had to get permission to go into his house and then was allowed to stay there for only a very short amount of time.

Conversations at the fire station always revolved around the same subject. " 'The home of the first couple rescued was shoved in the middle of the street,' " Pat Kelly read aloud from a pile of newspaper clippings. " 'The slide pushed one house into a tree, and then the debris kept crushing the house and pushing it around the tree and then pushing it into another house.' "

Pat stopped reading and turned to Jim. "That was your house," he said, "and I thought for sure it was going to take part of Doug's house off because it was leaning up against it."

Jim passed around paper cups imprinted with the Red Cross logo; then he uncorked the Merlot and poured it for everyone. The other house, he knew, belonged to Doug Spencer. Jim smiled, recalling how he'd helped Doug chainsaw a four-by-four-foot hole in the side of his garage. Doug had wanted to try to ease the press of mud against the walls by channeling it through the hole and down the hill. "We were there chainsawing, and after Doug left, one of the emergency workers said, 'What are you doing?' 'I'm making Doug happy,' I said."

Around the same time, another neighbor enlisted Jim's help in boarding up his house. "Against vandals?" Jim had asked. "No, he was boarding it up against the mud. This other guy was help-ing, and he said to me, 'What am I doing?' I said, 'You're making Kevin happy.'" For Jim, helping other people was a way of deal-ing with his own loss, which was total except for the house sign Pat Kelly had plucked from the flow of mud and the wonderful flying redwood bat that he himself managed to retrieve.

Jim was in the construction business. He was an accomplished carpenter, and he'd made the bat as a special present for his daughter, Kachina. It was a large fruit bat with hinged wings, and he'd hung it above the staircase on clear plastic wires so that it ac-tually seemed to be flying. "I want my bat," Kachina had said shortly after the slide; Jim's finding the bat quickly became some-thing close to an obsession.

The Kudroffs' house was the only dwelling that the slide had immediately destroyed, reducing it to a pile of splinters that were very nearly unrecognizable. But Jim methodically studied the pieces and turned them over in his mind as if they were pieces of a Rubik's cube. Somehow he found the dowel from which the bat had hung. Then he probed the rubble with his hands until he felt a smooth, rounded body and the tip of a great outspread wing. The other wing was missing. So he tried to figure out where it, too, might be hiding. He contemplated the smash of debris sev-eral times more. He dug through it again. When he finally found the wing, he felt as though he and the canyon had somehow set-tled a score. For several months, the bat occupied a place of honor in the Kudroffs' San Francisco living room.

The group at the firehouse talked for a couple of hours at least. Hee-Ran brought out a bottle of Korbel wine. Pat Kelly gunned up his motorcycle and raced up and down, followed by a franti-cally barking pair of Blue Heelers. The Kudroffs' children, Eric and Kachina, ran after them, squealing. Then the party broke up. A short time later, while Paula and the children sat with Hee-Ran in the small house she had rented near the fire station, Jim quietly disappeared. When he came back, he said that he had climbed to the top of the canyon and for the first time had seen the raw gash where the earth had given way. He said little else. He seemed somber somehow.

Paula, for her part, couldn't stop thinking about the house. "You know, I still can't figure out quite the mechanism of how it happened," she said. "The back is in the front; the front is in the back; part of the kitchen is here, part of it's there across the road. The living room's way over there, and the front porch is way down here, and it's like, gee, how did this all happen?"

ANATOMY OF A SLIDE

How did this all happen?

Well, El Niño certainly appeared to have had a lot to do with the timing of the landslide. At least this was the conclusion one could draw from a suite of computer runs undertaken by climate modelers at the National Oceanic and Atmospheric Administration some months later. What these scientists did was take the National Weather Service's medium-range computer model (the one that predicts the weather out to two weeks) and direct it to churn out a suite of "hindcasts" for California that first week in February.

In the first set of runs, the scientists gave the computer model misleading information about the condition of the tropical Pacific. That is, they supplied it with sea-surface temperatures that were based on "climatology," a kind of averaged-out smear of the seasonal pattern. In the second set of runs, however, they provided the model with the extreme-sea-surface temperatures that actually characterized the tropical Pacific at the time. The results seemed unequivocal: In the first instance, the model did not foresee any storms in California's immediate future, whereas in the second instance, it did.

But the storms that directly preceded the Rio Nido landslide also had idiosyncratic features that storm chaser Marty Ralph thought were extremely important. On a regional basis, he noted, the storm system that lashed California on February 5, and again on February 6, had not done as much damage as forecasters had feared. This was because, unlike its immediate predecessor, that storm system did not penetrate far inland, but reached peak strength offshore, then turned and raced off, rather swiftly, to the north. Even so, the clouds on the fringes of this immense system managed to dump an impressive amount of rain.

By this point, Marty had returned from his adventures in California to his government-issue office adjoining the University of Colorado campus. El Niño, he observed, is a global phenomenon, but people care about it only because of the ways in which it affects local patterns of temperature and rain. It was the fine scale of the interface between people and storms that Marty found fascinating. "The fine scale is where people live," he said. "It's what happens there that affects their lives."

Marty himself lived in a fine scale of creative clutter. Big rolls of meteorological maps stood in one corner of his office. Stacks of journals obliterated the desk. A cardboard box labeled Interesting Events sat on the floor. A leafy euphorbia, the kind known as a good-luck plant, lorded it over the Interesting Events. Marty had carted the plant with him from the University of Arizona, where he'd studied as an undergraduate, to UCLA, where he'd received his Ph.D., and from there to the National Oceanic and Atmospheric Administration's Environmental Technology Laboratory in Boulder, Colorado.

Like other fields of science, meteorology in the late twentieth century had splintered into a multitude of specialties and subspecialties. There were big-picture meteorologists who studied the flow of weather on a planetary scale; meteorologists who focused on a particular state or region; meteorologists who concentrated on particular types of storms, like hurricanes and extratropical cyclones. And then there were meteorologists, like Marty, who focused on structures, like the low-level jet, that were much smaller than the storms that spawned them.

Yet Marty was also linked, in a very direct way, to scientists who were struggling to understand the global climatological context in which the low-level jet was ultimately embedded. In his case, the link could be traced directly back to Jacob Bjerknes, who had done so much to lay the foundations for understanding both extratropical cyclones and El Niño. Indeed, Marty's thesis adviser at UCLA had personally known Bjerknes, who had come to UCLA from the Bergen School in Norway.

In 1998, Marty had flown through a passel of storms, but none, he said, left a more indelible impression than the two that arose during that first week of February. Both those storms, he observed, had boasted low-level jets of incredible ferocity. Out over the ocean, at a height of 3,000 feet, he and his colleagues had measured winds whose speed ranged between 80 and 100 knots. They had also measured dew points so high (the dew point is the temperature at which water vapor condenses into rain) that it often seemed as though they were riding through a summertime thunderstorm.

Always the P-3 had penetrated the low-level jet through the cold front on the southeastern edge of the storm, Marty recalled, "and every time, it was like stepping out into Florida; it was a muggy, sultry feeling." The winds feeding the low-level jet appeared to be coming from extremely far south. "It's not conclusive," Marty mused, "but it's certainly suggestive that these storms were sucking warm, moist air straight out of the tropics."

The low-level jet occupies a kind of netherworld between the clouds and the ocean. It starts just above sea level and parks at around 3,000 feet, giving it an elevation range comparable to the low-lying mountains that extend from the southern part of the state all the way to the Oregon border. "These are not mountains like Rainier or Everest," said Marty. "They're just little blips, but as far as the atmosphere is concerned, they're big." Indeed, it's because of these little blips that the conversion from water vapor in the air to rain on the ground is so incredibly efficient. As soon as the low-level jet slams into a steep hillside, the air is forced up and into a colder region of the atmosphere where the dew point is much lower.

From instrument readings taken during the P-3 fly-throughs, Marty continued, it appeared the warm air surging through the low-level jet had crashed into a heavy mass of cold air trapped on the windward side of the coastal mountains. What occurred, in other words, was a collision between warm- and cold-air masses that produced the equivalent of a storm within a storm. As the warm air in the low-level jet was forced up and over the cold air at the base of the coastal mountains, then over the mountains themselves, the temperature in the jet plunged below the dew point so fast that as soon as clouds formed, the rain started to fall.

In order to produce rain, droplets in clouds (in big convective storms these "droplets" often start out as ice crystals) have to reach a certain size, and the way cloud droplets grow is by merging with other droplets that they hit on the way down. In the case of the storm that immediately preceded the Rio Nido landslide, the droplets formed so close to the ground that they did not have a chance to grow large, but bombarded Upper Canyon Three while remaining very small. The low-level jet was like a powerful water gun, only in this case it was filled with a fine, misty drizzle.

This explained a small mystery associated with the landslide: how a rain gauge at Cazadero Creek could record 9 inches in the space of forty hours, even though, to many of those who lived nearby, it did not seem to be raining all that hard. It did not explain the larger mystery, however. If El Niño had cocked the gun in February 1998 and the low-level jet had pulled the trigger, then what was it that loaded the gun and left it at the top of Upper Canyon Three?

THE RUGGED TERRAIN along the California coast is notoriously prone to landslides, and unbeknownst to those who lived in Rio Nido, the forested slope above Upper Canyon Three was becoming increasingly unstable. In fact, part of the slope may have started downhill thousands of years ago, creating a structural defect akin to a bone fracture that had only healed partway. At least this was the possibility that occurred to William McCormick, a geological consultant at the Santa Rosa, California, firm of Klein-felder Inc., as he approached the spot where a staircased terrace built of sandstone had just weeks ago given way like the rungs of a rain-rotted ladder.

This was landslide-prone country, said McCormick, who was the geologist then advising Sonoma County. There was no question, for example, that the bowl-shaped depression on the other side of the canyon had been carved by a landslide. To a geologically trained eye, the depression bore eloquent witness to an ancient tumbling of rock and soil some ten times larger than the 1998 slide. Whether that landslide had come down all at once or in incremental bits and pieces was hard to say, for until 1806, when Russians in search of fur seals sailed down from Sitka, Alaska, this section of coastal California was inhabited only by small roving bands of Native Americans. In those days, a landslide could roar down Upper Canyon Three without smashing into any houses, and a big tree could fall and make no sound for lack of people to hear it.

Those who settled in the vicinity of Rio Nido in the nineteenth century were certainly aware of landslides. In 1872, for example, a landslide cut off the railroad that linked the lumber town of Guerneville to the Korbel brothers' property farther east. But compared to floods and earthquakes, landslides seemed too small in scale to warrant a second thought. In truth, landslides were a negligible threat until recently; in the twentieth century, however, as a consequence of the widespread movement of people into steep terrain, landslides have come into their own as an extremely important natural hazard. In the United States alone, landslides are thought to cost an average of twenty-five lives and a billion dollars in property damage every year.

Landslides can be triggered by a variety of mechanisms, including volcanic eruptions and earthquakes. In 1920, for example, an earthquake in China's Gansu Province set up landslides that are thought to have killed one hundred thousand people. In 1970 an earthquake off the South American coast sent a landslide hurtling off the peak of Huascarán, the highest mountain in Peru, killing eighteen thousand people. In 1980, the explosion of Mount St. Helens triggered an avalanche of mud and debris that buried highways, destroyed bridges and houses, and dammed rivers to create new lakes.

But the most common cause of landslides is an overwhelming surfeit of rain, and in California, in the winter of 1998, rain was definitely the culprit. Landslides occurred inland, in the craggy hills and canyons of the coastal mountains, and along the coast itself as heavy rain and high waves staged pincer attacks on scenic, steep-sided cliffs. In Pacifica, just south of San Francisco, residents of houses built atop one such cliff found that the gravel and sand underlying their yards had melted away like powdered cocoa in a pot of boiling water. Within a few weeks' time, they ceded to the ocean some 30 feet of prime beachfront property, more than fifty times the loss associated with an average winter.

In Los Angeles a mud slide rolled into a house like a big bulldozer and pushed it into the backyard swimming pool. The hole the landslide tore from the hillside left four other homes teeter-tottering on the edge of a yawning abyss. In Clearlake Oaks, two houses slipped and skidded downslope, threatening to fall onto Highway 20. In Ventura, mud slides ruptured oil pipelines and dammed a canyon to create a lake that menaced some fifty houses downstream. In La Honda, some two-dozen houses began a spooky slow-motion fall as the ground beneath them began to slink downhill like Silly Putty. And then there was Rio Nido.

Like much of coastal California, the area around Rio Nido sits not so firmly astride a jumble of rock that includes the infamous Franciscan formation, a crazy quilt of sandstone, shale, and chert interspersed with blocks of blue schist and serpentinite. Sandstone, shale, and chert are sedimentary rocks; they were formed as layers of sand and mud and silica at the bottom of the ocean were pressed together by the weight of the overlying water. Blue schist and serpentinite, on the other hand, are metamorphic down hillsides at speeds approaching 30 miles per hour. These landslides, known as debris flows, generally involved only the topmost 3 feet or so of soil and rock, and they usually cut a swath that was not much more than 100 feet wide. Even so, debris flows ranked as the most dangerous of all types of landslides. When they collided with structures like highways and houses, they did so with the force of an oncoming truck.

Earlier in the week of the Rio Nido slide, Loma Mar, California, a community in the Santa Cruz Mountains, was hit by not one but several debris flows. Forty-six-year-old Paul Henzel, the head of a local water company, died after a barrage of rocks and trees and mud swept through his bedroom, carrying him with it. Two neighbors found Henzel buried in waist-deep mud, seriously injured but still alive. They dug him out, carried him to the general store below, and administered what first aid they could. Then they waited for help.

The problem was, emergency helicopters were unable to land right away because of high winds and low visibility, and ambulances could not reach the area because of the flooding triggered by the slides. Along with the debris flows came dozens of trees, which tumbled into the narrow channel of the Pescadero River, then jammed beneath a bridge, forming an impromptu dam. Paul Henzel died, and whenever the residents of Rio Nido thought about what had happened to him, they shuddered.

The Loma Mar landslides were simple debris flows. The Rio Nido landslide, by contrast, was more complicated, a hybrid landslide that combined a deep-seated, slow-moving slumping of earth with two debris flows. The chain of events occurred as follows: First, a big rectangular block of earth and rock, which measured 400 feet by 200 feet by 50 feet, lost its purchase on the slope, sliding forward like a plate slipping out of a cupboard. Then, with nothing but air to support its multi-ton weight, the outermost part of the block abruptly fell away. It was this falling away that powered the debris flows that followed.

One of the debris flows raced through a ravine that fed into Upper Canyon Three from the north; it carried so many Douglas rocks; they were forged in the geophysical pressure cooker that lurks beneath the earth's crust.

The Franciscan formation was forced upward shortly after the demise of the dinosaurs as the North American continent collided with a much thinner wedge of oceanic crust on its far western flank, compressing and crumpling the rock at the margins. As a result, the steep-sided canyons and rough-hewn mountains that crinkle the California coastline contain untold cracks and fissures into which air and water continuously seep, breaking rock down into grains of sand and silt and minerals such as feldspar. This relentless process is known as weathering, and its end product is soil.

By the winter of 1998, the steep slopes above Upper Canyon Three had accumulated a thinnish topping of soil atop a much thicker layer of deeply weathered rock. Every time it rained, the interstices in the soil and the rock filled with water like the pores of a sponge, exerting an outward pressure that forced the surrounding grains of silt, sand, and clay apart. If the terrain happened to be flat, this increase in "pore pressure," which was what hydrologists called it, merely made the ground underfoot inconveniently soggy. But if the terrain happened to be steep, the entire hillside might lose its structural integrity and start to tumble. This is what happened up and down the California coast during the El Niño winter of 1997–98 as storm after storm rolled in from the ocean.

In many ways, of course, these storms were beneficent forces. The rainfall they generated refilled reservoirs and recharged rivers and streams and aquifers. The problem was, there were too many storms spaced too close together, and their gifts of precipitation were too generous. As a result, the greater San Francisco Bay area recorded two and a half times the normal amount of rain. Not by accident, this same area also recorded nearly five hundred landslides that caused significant damage.

Many were slow-moving landslides that clenched houses in muddy vise grips, wrenching them apart over the course of days and weeks. But others were fast-moving landslides that raced firs that it created a big logjam, 20 or so feet high, at the end of the road. The other funneled into the winter creek above the La-Combes' house and plunged downhill some 1,500 feet. It was this debris flow that had terrified Phyllis and Gary as they sat at their kitchen table and that had slammed into the Kudroffs' house, crushing it against a tree.

For the residents of Upper Canyon Three, the problem was that the danger did not end on that February night, as geological consultant William McCormick informed officials of Sonoma County in a tersely worded preliminary report dated February 18:

> Since initial failure of the upper block, a crack approximately 125 feet long has developed north of the existing head scarp (uppermost shear plane), along a portion of the hillside that has not yet failed. Within the last week, this crack has widened to 6 inches across and has dropped approximately 7 inches vertically. Movement along this crack indicates that additional portions of the ridgeline are moving and have potential to develop into subsequent block and debris flow failures. During the last week, the head scarp (back face) has experienced localized failures. . . . Open tension cracks have developed in the outer, eastern portion of the landslide block. These cracks have increased in width and depth throughout the week, resulting in cracks up to 10 feet wide and 7 feet deep. The outer face of the block continues to deteriorate, and sections sporadically fall off in soil/rock masses. . . . Debris from these failures tends to accumulate at mid-slope elevations and is subsequently transported down slope by storm erosion.

What concerned McCormick was the potential for a much wider disaster, a catastrophic failure that would not only overwhelm the remainder of Upper Canyon Three but also threaten houses lower down by damming the creek Phyllis LaCombe called White Water and turning it into a lake.

Such a disaster could be touched off by another exceptional

spate of rain or by a moderately strong earthquake, for that matter. The San Andreas fault runs just to the west of Rio Nido, the Healdsburg fault to the north. It would be an unlikely but not impossible coincidence, geologists agreed, if a temblor should occur just when soils were supersaturated with rain. To monitor the constantly changing conditions, scientists from the U.S. Geological Survey installed a battery of instruments at the top of Upper Canyon Three, including instruments called extensometers that measure movement. The instruments were sunk into plastic plumbing pipes and flagged with pink and orange ribbons that fluttered like flags.

Despite the bright colors, the scene where the earth had so recently given way was strangely disconcerting. A wide slope of once-upright Douglas firs slanted sideways, as if bending in a strong wind, but it was not the trees that had adopted this oddly angled stance. Rather, the ground itself had sharply tilted so that what had been a downhill slope now tipped uphill, and near the edge of this broken block of earth, tree roots protruded from a gold-brown amalgam riven with fractures of fearsome size.

Dry, this clay-rich admixture was hard as brick but a lot more brittle; pressed between thumb and forefinger, it crumbled like a cookie. Wet, it underwent a radical change, turning into a slurry the color of café au lait and the texture of newly mixed concrete. Because of his background in construction, Jim Kudroff was familiar with concrete. "I pour it all the time," he said. "The hardest part is trying to form it against a hill, because all that pressure is going to come at you, just like this did. I loved it when one guy stood up at a meeting and said, 'Why don't we just build a retaining wall up there to hold it?'"

"Sure." Jim laughed to himself. The partially tipped bucket at the top of Upper Canyon Three, he knew, held the equivalent of 250,000 cubic yards of concrete mix, enough mix to fill a multistory building the size of a football field, enough to fill several San Francisco four-flats and their backyards, enough to fill the elementary school his daughter, Kachina, attended. Indeed, the amount of material suspended over Rio Nido weighed as much as thirty thousand full-grown elephants, and should it opt to rush downhill all at once, no retaining wall in the world would be strong enough to stop it.

In the end, this much-feared catastrophe did not occur. Toward the end of March, the storms started to taper off. By midsummer, the soggy earth had turned solid again. Even so, like the boulder the gods are said to have positioned above the head of the mythical character Sisyphus, the big block of earth hovering above Upper Canyon Three continued to disturb the sleep of those with houses below. There was a temporal delay built into the play of rain upon landscape, it seemed, so that the danger to human welfare posed by storms could linger for a surprisingly long time.

CHAPTER **SEVEN**

As long as the Rachel *was in view, she was seen to yaw hither and thither, starboard and larboard. But by her winding, woeful way, you plainly saw that this ship still remained without comfort. She was Rachel weeping for her children, because they were not.*

HERMAN MELVILLE, *Moby Dick*

OUT OF AFRICA

In early February 1998, C. J. Peters, a short, squarish man with a salt-and-pepper mustache, boarded the first of a series of planes that took him from Atlanta, Georgia, to Garissa, Kenya, the epicenter of an epidemic that had just blasted through northern Kenya and southern Somalia like a biological storm. There he and his team from the U.S. Centers for Disease Control and Prevention joined other experts on infectious disease gathered together by the World Health Organization, and for the next several weeks they worked out of a safari-style camp, documenting the most extensive outbreak of Rift Valley fever on record.

There were other diseases mingled in, so it was hard to come up with a precise number. But in all, it was thought, up to ninety thousand people in this desolate, semidesert region may have been stricken by this mosquito-borne illness, and of these as many as one thousand may have died. Even more devastating in the long term was the economic blow that the simple pastoralists and nomadic herdsmen of the region sustained. "These people did not have bank accounts at Chase Manhattan or equity in the stock market," observed Ali Khan, a senior epidemiologist on the team C.J. hastily assembled. "All the wealth they had in the world was walking around on the hoof." In just a few months, they lost some 20 percent of their cattle and camels and some 70 percent of their goats and sheep.

The Rift Valley fever outbreak, it should be noted, was just the latest in a series of tribulations endured by the people who inhabit the politically and economically stressed region that straddles the border between Kenya and perpetually war ravaged Somalia. It struck just as they were beginning to recover from a drought so severe that once-reliable wells pumped up empty gushes of air and local watering holes dried into bowls of cracked clay. The parching closely paralleled the evolution of the La Niña that began in 1995 and reached peak intensity in 1996. That was the year when the so-called long rains taunted this vast, unfenced rangeland with their extreme parsimony, and the short rains were so short that they might as well not have come at all.

Like the sun-baked ground underfoot, family units developed cracks and fissures. Men and older boys stayed in the countryside with their herds, congregating around the few remaining sources of grass and water, while their wives and young children sought refuge in the dusty towns strung out along the banks of receding rivers like terra-cotta beads. In order to reach these little outposts of semisecurity, small groups of women, babes on their backs, water jugs on their heads, often walked a hundred miles or more across terrain that was as treeless and grassless as a moonscape.

Then, in October 1997, the drought ended, and the deluge began. Virtually every afternoon, towering clouds, dark with rain, massed along the East African coast. From Tanzania in the south to Ethiopia in the north, rain fell. It fell on the short grasses nibbled by wildebeests and zebra in the Serengeti Plain. It fell on the cultivated fields surrounding Lake Victoria, on the papyrus-fringed marshes of Lakes Naivasha and Nakuru, on the bamboo forests of Mount Kenya, whose streams feed the Tana and Ewaso Ngiro Rivers, and on Ethiopia's Mendebo Mountains, source waters of the Juba and Shabelle Rivers that flow through southern Somalia.

At first, the rains were greeted with relief. All too soon, however, the trickles of water turned into torrents. Bloated by rain, rivers repeatedly burst through the confining seams of their banks, ruining crops and sweeping away bridges and highways. For a while, the Juba and Shabelle merged, turning tens of thousands of acres of Somalia's best farmland into a polluted inland sea. To escape the rising waters, people living alongside climbed onto roofs of shaky thatch or huddled with their animals on little islets of higher ground. Where there was no ground, they climbed up acacia trees with thorny branches, though most of the trees that they might have climbed had long ago been felled for firewood. Several thousand people drowned in the floods, and hundreds of thousands were forced from their homes.

Soon came the first reports of cholera and malaria in the flood-ravaged areas. Given the severity of the flooding, public-health officials had expected outbreaks of those water- and mosquito-borne diseases. Then, in late December, rumors of a deadly bleeding disease started wending their way out of the hinterland. On December 23, 1997, Louise Martin, a Nairobi-based veterinarian trained by the Centers for Disease Control, helicoptered into Kenya's waterlogged north as part of a World Health Organization team. Upon her return, she posted an alarming E-mail about the situation in Masabubu, a village of twelve hundred located on the Tana River to the south of Garissa.

In the two weeks prior to Christmas, she reported, sixteen residents of Masabubu had died, many of them members of the same families. One man, for example, reported losing his wife just before Christmas. She was twenty-eight years old, wrote Martin, "became sick about one week ago, was sick for twenty-four hours with severe headache, back pain, fever, vomited normally about three times, then vomited blood, diarrhea'd blood, and lived for only twelve more hours."

And that wasn't all. The man's sister-in-law (his wife's sister) had died in a similar fashion some days before, as had a male cousin, wrote Martin.

These are nomadic people, and they were beginning to move on. [The cousin] complained of fever on the first night, and said he couldn't walk and so he was left behind with two men to help him. They report that after a short time he vomited blood, diarrhea'd blood, was bleeding from the nose and mouth and then died. From signs to death, about 12–24 hours. This man had sick and dying sheep and goats, no knowledge of his eating them. . . .

Within 20 paces of the school in the village, I counted 18 goat carcasses from goats that had died before they could be slaughtered, and were dragged out to the field away from houses. The villagers say the hyenas and the vultures don't even eat them, although I saw a dog chomping down pretty well. Did I mention that this is also a big problem in Somalia as well. . . . There are camels dying in serious numbers with the same sort of symptoms, and the people eat the dying camels.

In another village she visited, Martin found a fourteen-year-old girl who was on the verge of dying. The family had killed and eaten a sick goat the week before, she observed. The girl was the only one to have eaten the liver. In the same village, Martin found another fourteen-year-old with pus-filled blisters all over her hands; she had been helping an older sister butcher livestock.

From Martin's reports, C. J. Peters very quickly developed a strong suspicion that the source of this widespread malaise was Rift Valley fever, a disease that chiefly affects domestic animals but can spread, sometimes explosively, to humans. As he well knew, cows, camels, sheep, and goats all "catch" Rift Valley fever from virus-laden mosquitoes, and people frequently become infected that way, too. But people can also become infected by coming into close contact with sick animals—by slaughtering them, by drinking their blood or milk, or by eating their meat, especially meat that is undercooked or raw.

What was at first just an educated guess hardened into conviction after blood specimens that Martin and her colleagues collected from sick people were shipped from Kenya to the National Institute of Virology in Sandringham, South Africa, and C. J. Peters's laboratory in Atlanta. After screening the specimens for the presence of numerous deadly pathogens, including the Ebola, Machupo, and Lassa viruses, both labs very quickly converged on the same answer. The mystery disease was indeed Rift Valley fever.

C.J. had long been fascinated by Rift Valley fever. He first tangled with the disease in 1977, shortly after he signed on with the U.S. Army Medical Research Institute in Fort Detrick, Maryland. That was the year when his bosses, concerned about the virus's potential as an agent of biological warfare, sent him to investigate a very large outbreak in Egypt that, for the first time, had resulted in hundreds, perhaps thousands, of human fatalities. Up to that point, C.J. had never heard of Rift Valley fever, but he very quickly read up on what was then known about it, beginning with a paper published a half-century earlier in the *Journal of Pathology*.

In that paper, R. Daubney and J. R. Hudson, two veterinarians from Nairobi, described the 1930 outbreak of a mysterious disease on a sheep farm in Kenya's scenic Rift Valley. This dramatic gash in the earth's crust, framed by craggy escarpments, marks a 4,000-mile-long corridor of subterranean stress where the African continent and the Arabian peninsula are slowly pulling apart. Among the legacies of this tectonic event are the Red Sea, the Serengeti Plain, and a double-stranded necklace of jewel-like lakes.

The affected farm encompassed some thirty thousand acres of grass and bushland on the edge of one of the most beautiful of these lakes, Lake Naivasha. The first signs of trouble had shown up in June 1930, Daubney and Hudson recounted, and had coincided with several months of record rain. First the pregnant ewes on the farm began aborting their young. Soon afterward, newborn lambs started to die off. A short time later, many full-grown ewes were dying as well. By the end of July, the farm had lost thirty-five hundred lambs and fifteen hundred ewes.

As Daubney and Hudson soon learned, however, the outbreak of disease was not confined to sheep. Cows were affected by it, as were humans. Indeed, Daubney and Hudson themselves wound up contracting the disease. "The onset," they dispassionately reported, "was characterized by a very brief period of general malaise, followed rapidly by sharp rigors and headache. The temperature rose to about 103 degrees, and the face was brightly flushed; in from three to six hours after the rigors had passed off, pains developed in or near the joints extending from the base of the skull to the extremities . . . fever persisted for a period of from 12 to 36 hours, and the pains gradually disappeared within about four days."

Over the course of several months, Daubney and Hudson methodically tracked down the source of the outbreak, which they established had affected not just the sheep on the Lake Naivasha farm but "almost every native engaged in herding [the] sheep." First, they took serum from affected animals and used it to inoculate healthy sheep; when those animals sickened, they deduced that they were dealing with an infectious agent. Then, in an experiment that would be considered unethical today, a native tribesman was inoculated with the serum as well. Not surprisingly, he, too, sickened, and serum prepared from his blood proved similarly infectious to animals.

Initially, Daubney and Hudson speculated that they might be dealing with a bacterium in the *Salmonella* genus, but when they passed serum from diseased animals through filters fine enough to remove bacterial particles, the serum lost none of its infectious power. Like yellow fever and dengue, they concluded, this disease must be caused by something smaller than a bacterium—a virus, in other words. And like yellow fever and dengue, they went on to speculate, it was likely to be transmitted by mosquitoes.

During 1930, the year of the outbreak, they noted, the region around Lake Naivasha recorded 45 inches of rain, 20 inches more than normal, and mosquitoes were "particularly numerous." Among other things, transmission of disease abruptly stopped when sheep were moved from the farm, which was located at an elevation of 5,500–6,000 feet, to an altitude of 7,000 or more feet, where the cooler air served as a natural insect barrier. It also stopped when sheep were placed in cages draped with netting that protected them from mosquito bites.

It took scientists nearly eighty years more to understand the link to climate that had been implicit in that outbreak, for like 1997–98, 1930–31 had been an El Niño year. No less than landslides and floods, Rift Valley fever outbreaks were manifestations of an underlying meteorological teleconnection. While El Niños did not invariably trigger such outbreaks in East Africa, they did so more often than one would expect by chance. Moreover, the correlation strengthened when scientists applied a complicated metric that included not only the swings in atmospheric pressure and sea-surface temperatures in the tropical Pacific but in the Indian Ocean as well. When climatological conditions in both oceans were fatefully aligned, the chances of exceptional rains in East Africa—and hence, outbreaks of Rift Valley fever—rose.

WITH EACH NEW El Niño, it seemed, scientists learned of yet another organism whose biological fortunes oscillated to its strikingly idiosyncratic drumbeat. These oscillations took the form of sudden population crashes that approached precariously close to the brink of extinction, followed by increases in population so prodigious that they tested credulity. Indeed, when plotted on a piece of graph paper, the ups and downs sometimes mirrored the swings between El Niño and La Niña so perfectly that it almost seemed as though the biosphere were a dancer swaying to some exotic form of music.

Consider, for example, *Nolana*, a genus of pretty flowering plants that in early 1998 produced an ephemeral explosion of color in the coastal desert of Peru. Unlike cacti and succulents,

Nolana requires ample amounts of moisture in order to grow. The seeds it produces, however, have the capacity to remain dormant for years. Only when El Niño comes to awaken them with sloppy kisses of rain do these sleeping beauties put forth roots and leaves and burst into bloom, temporarily transforming barren hillocks into undulating carpets of blue.

It was plants like *Nolana* that Alfred F. Sears had in mind when he described the magical transformation wrought by the 1891 El Niño to fellow members of the American Geographical Society. As he put it,

> Once in five to seven years there comes a marvelous change. . . . A shower, sometimes two or three, wanders away from the mountains and is poured over the arid land; then what wealth of beauty leaps under the wand of nature! The hitherto lifeless earth springs into being; grass and flowering plants appear on every hand, grown to the height of a horse's head. . . .
>
> Whence comes all this brilliant life? How has it slept in the ground during these years? It lasts a week and has been known to live nearly a month; cattle and great herds of goats wander out of the inundated valleys across the living field and revel in the boundless joy.

The magic wand El Niño waved over Peru's coastal desert also brought hordes of frogs and rats, crickets and spiders, dragonflies and mosquitoes. During the 1925 El Niño, for example, a gentleman traveling by rail from Paita to Piura was struck by the many newly formed lakes that provided mosquitoes with ideal breeding habitats. "Malaria," he wrote, "is now rife in Catacaos and also in the Chira River towns. Rheumatic complaints are common, and it is asserted that dengue, or breakbone fever, has appeared in Paita."

As it turned out, it wasn't just the rain-drenched coast of Peru that El Niño plagued with malaria and dengue; somewhat counterintuitively, as public-health experts later established, outbreaks of these diseases were also common in areas that El Niño afflicted with drought. Thus, in coastal Peru, the *Anopheles* mosquitoes, which carry malaria, hatched out in temporary lakes in the desert, whereas in the Colombian highlands, they flourished when a dearth of precipitation turned fast-flowing rivers into a series of stagnant ponds.

The same was true for *Aedes aegypti*, the mosquito that carries both the yellow fever and dengue fever viruses. This striped-legged creature could breed virtually anywhere it could find a smidgen of standing water: in abandoned tires, tree holes, septic tanks, roof gutters, cemetery urns, old stumps, flowerpots, tin cans, buckets and cisterns. Wet spells that filled these ubiquitous receptacles suited *Aedes aegypti*, but so did dry spells that en-couraged people to store water in uncovered containers. It was no accident, in other words, that between January and May 1998, the El Niño–triggered drought in Indonesia produced a very high number of dengue cases: 32,666, of which 774 were fatal.

It wasn't until the 1980s that scientists finally zeroed in on an-other type of *Aedes* mosquito whose life cycle, in retrospect, al-most seemed to have been designed with El Niños in mind. These were the floodwater *Aedes* that carry Rift Valley fever, and they had perfected a *Nolana*-type strategy for dealing with sharp swings between rain and drought. These mosquitoes lay their eggs not in standing water but in the mud of recently flooded depres-sions. As the mud dries, these eggs then enter a state of suspended animation that lasts until the next set of torrential rains comes along to flood the depression again.

At that point, the eggs very quickly hatch, and the progression through the life cycle goes into fast-forward. First to emerge are wriggly larvae that float upside down and breathe through little tubes like miniature snorkelers. After molting several times, the larvae fatten into free-swimming pupae ("tumblers," entomolo-gists call these vaguely shrimplike creatures), and then, in a breathtaking metamorphosis, the tumblers shed their shells to re-veal full-grown insects with gauzy wings, long, delicate legs, and needle-sharp proboscises. In late 1997, it was just such a hatch-out of floodwater *Aedes*, probably *Aedes macintoshi*, that set the stage for the Rift Valley fever outbreak in northern Kenya and southern Somalia.

The connection between Rift Valley fever outbreaks and intense rains had intrigued C. J. Peters for fifteen years, ever since he'd puzzled over the pattern of Rift Valley fever antibodies that he'd found in blood samples collected from people living in the Central African Republic. Those living in the rain forest, the wettest zone, had the lowest antibody counts; those living in grassland savannas and dry forest had counts that were sometimes half again higher. "That's when we realized there was no relationship to the total amount of rain; if anything, there was almost a reverse relationship. So that's when we began to think, maybe it's not the rainfall per se; maybe it's the flooding."

The idea that flooding was important strengthened in the 1980s after a South African veterinarian named Robert Swanepoel meticulously described the ecological setting of the episodic Rift Valley fever outbreaks plaguing Zimbabwean cattle ranchers. The outbreaks, he found, tended to cluster around hydrologic structures known as *vlies,* which were typically located at the edge of rivers and seasonally flooded woodlands. The *vlies* resembled nothing so much as stopperless bathtubs. In this case, the cracks in the underlying rock served as the drains through which water flowed into a nearby stream or river. The *vlies* could not flood, in other words, until enough rain fell to raise the surrounding water table.

Soon afterward, a series of field experiments undertaken by U.S. Army scientists working in Kenya produced a compelling picture of how torrential rains plus *vlie*-like structures known as *dambos* in Swahili could power a Rift Valley fever outbreak. During the 1982–83 El Niño, for example, these scientists collected mosquito larvae and pupae from flooded *dambos* in the central part of Kenya. They did the same thing in 1984 when they forced a similar hatch-out by pumping water out of a nearby river to flood a dried-up *dambo* in the same vicinity. They then took the larvae and pupae they harvested in this fashion to their laboratory, reared them into adult mosquitoes, and tested them for the presence of the Rift Valley fever virus before they'd had a chance to feed.

What they found was interesting. The male mosquitoes (which feed only on nectar) carried the virus as well as the bloodthirsty females. Unlike the mosquitoes that carry malaria and dengue, in other words, female floodwater *Aedes* did not have to bite an infected animal or person in order to pick up the Rift Valley fever virus. They were born bad, as it were, having acquired the infection from their mothers, so that the very first time they bit a warm-blooded animal, they passed the disease-causing virus along. Most years, the infected *Aedes* did this at very low levels, if indeed they managed to hatch out at all. In El Niño years, however, there was often an additional, amplifying factor.

For by themselves, it seemed, these floodwater *Aedes* were not capable of creating biological mayhem. They needed accomplices, and accomplices, in the guise of other mosquitoes, were what extended periods of heavy rainfall provided. The floodwater *Aedes* hatched out first. Then other types of mosquitoes hatched out. Some of these mosquitoes bit the same warm-blooded animals that the floodwater *Aedes* had infected with Rift Valley fever, and they became infected in turn. The second time these mosquitoes bit, they injected their victims with tiny droplets of virus-contaminated saliva. In this way, the circle of infection geometrically expanded.

There was a delay, in other words, between the hatch-out of the virus-carrying *Aedes* and the hatch-out of other types of mosquitoes that helped spread the virus. C.J. felt that was important. It meant that the Rift Valley fever outbreak in northern Kenya and southern Somalia started only with the flooding of the *dambos* along the Tana, the Ewaso Ngiro, and the Juba and Shabelle Rivers. It meant that even in October 1997, after the rains started to fall, there was still time for public-health officials to have identified the areas at highest risk and taken some precautionary measures.

Indeed, as he contemplated the suffering that Rift Valley fever had caused around Garissa, C.J. started to wonder if scientists might have more effectively combined what they had learned about the virus and the mosquitoes that carried it with what they had learned about El Niño. He wondered if they might have used that knowledge to forewarn people in areas at greatest risk and even, perhaps, prompt action that would have prevented the epidemic from ever occurring.

OF MICE AND MEN

In 1992, at the age of fifty-three, C. J. Peters had been only peripherally aware of El Niño. He had just become the head of the Special Pathogens Branch at the Centers for Disease Control. A short time later, he and his wife, Susan, moved into a modern house in the Atlanta suburb of Stone Mountain, Georgia, where his collection of Central Asian carpets incongruously vied with her collection of ladies' hats. In the backyard was a small pond filled with water lilies. It was a low-maintenance garden, C.J. explained, ideally suited for someone who spent the greater part of his life hopping on and off airplanes.

By this point, C.J. had become something of a popular icon, having played cat and mouse all over the world with viruses, like the Ebola virus, that were both lethal and, thankfully, very uncommon. Many of these viruses were the causes of so-called hemorrhagic fevers that triggered unstoppable bleeding. Sometimes the viruses themselves punched tiny holes in the linings of blood-vessel walls; sometimes they stimulated the immune system to release potent chemicals that had the same effect. Sometimes they caused the formation of microscopic clots that impeded the flow of blood and caused blood vessels to swell. Sometimes they prevented clotting altogether.

In 1971, C.J. had played a role in the investigation of an outbreak of hemorrhagic fever in Cochabamba, Bolivia, which had cut a lethal swath through the medical staff at the Catholic-run Seton Hospital. It was caused by the Machupo virus, named after a river that feeds into the drainage basin of the Amazon. His familiarity with the virus explained why, in early 1993, he found himself back in Bolivia trying to assess whether or not the virus that caused this deadly disease was preparing to reemerge. A new case—the first new case in nearly twenty years—had triggered concern, and so officials in the Bolivian government called in the two best experts they knew. C.J. was one. The other was University of New Mexico biologist Terry Yates, who had spent years roaming the most remote corners of Bolivia trapping mammals of all kinds, including the rodents that carry Machupo. At the time, Terry was not interested in disease at all; he was intent on assembling an evolutionary history of the mammals that populate this part of the world. He and C.J. met for the first time in the dizzyingly high capital of La Paz, sized each other up, then headed out together into the lowlands of the Beni, where the Amazonian rain forest abuts the foothills of the Andes.

C.J. and Terry soon found they had a lot in common. Both came from small-town backgrounds. C.J. had grown up in West Texas; Terry, in western Kentucky. From a very early age, both had been drawn to biology, and as they splashed through rivers on horseback and rode bicycles from one village to the next, they started talking. "His interest was the virus," recalled Terry, "and my interest was the host." The Machupo virus was a soccer ball–shaped curl of RNA that appeared to be transmitted by rodents directly without the help of biological syringes like mosquitoes, ticks, or fleas. It was a relative of the African virus that caused the hemorrhagic disease known as Lassa fever, C.J. mused. Perhaps both viruses radiated out from a common ancestor on the ancient supercontinent of Gondwanaland, which, before its penultimate breakup some fifty million years ago, included both Africa and South America.

C.J. was much drawn to the idea that disease-causing microbes were, for the most part, extremely old. Indeed, virtually all the pathogens that afflict humans today, he noted, had somehow bootstrapped their way up from the animal world. The virus that causes yellow fever, for example, is thought to have taken up residence in monkeys before spreading to humans. In similar fashion, the parasites responsible for malaria appear to have jumped to humans from rodents and birds and other primates. The sandfly-borne parasite that causes the disfiguring and sometimes fatal disease known as leishmaniasis, for its part, may well have included among its vertebrate prey the now-extinct dinosaurs.

Terry Yates liked the idea that the Machupo virus was old, but he didn't buy the notion that it was all that old. In South America, he pointed out to C.J., Machupo appeared to be confined to a single family of rodents that were thought to have crossed from Eurasia to North America via the Bering Strait land bridge when sea levels plunged during an epoch of glacial advance. Very slowly, as Terry envisioned it, members of this rodent family drifted south, eventually wending their way into South America over the Isthmus of Panama.

These rodents were grassland creatures, Terry observed, and the drier climates that marked glacial high-stands would have aided their spread by favoring grasslands over forests. In Latin, the family name of the interlopers was Muridae, and *Calomys callosus,* the mouselike creature that carried the Machupo virus, was a member of the clan. As Terry imagined it, the ancestors of *Calomys* and all the other New World murids followed grassland corridors down from Canada, across Central America, and from there, into South America. "It was at this point," he said to C.J., "that the wave of invaders from the north spread out across the South American landscape and started diversifying." And so, Terry thought, did the viruses the invaders carried.

The link between the Machupo virus and *Calomys callosus* had been pinned down in the wake of an epidemic that struck the little town of San Joaquin in 1963. Over the course of two years, fully a quarter of the town's twenty-five hundred residents came down with Bolivian hemorrhagic fever, the disease caused by the virus, and of these, 113 died. When scientists from the Middle America Research Unit, a now-defunct Panama City–based branch of the U.S. National Institutes of Health, went to San Joaquin to investigate the outbreak, they had to be evacuated after three members of the team themselves came down with the illness. Among them was the leader of the team, Karl Johnson.

Up to that point, rodents had been implicated as carriers of numerous bacterial pathogens, like plague and typhus, but only one virus that caused human disease. Johnson, however, became convinced that rodents were the conduits for Machupo, and after recovering from his own bout with Bolivian hemorrhagic fever, he mustered the evidence to prove it. He divided the town of San Joaquin in two. On one side of town, he instructed residents to start trapping rodents in early May. The number of new cases of the disease soon dropped. Then he encouraged residents on the other side of town to start a trapping program as well. In a very short time, the epidemic ended.

What happened in San Joaquin? Sporadic cases of the disease had been noticed for some time, no doubt because loggers and farmers who ventured into more remote areas sometimes came into contact with infected *Calomys*. But in 1963—due to some sort of environmental stress, increased competition for food, perhaps—large numbers of these small brown rodents scurried out of the countryside and into the houses of San Joaquin. Perhaps there was some link to fluctuations in rainfall; perhaps not. Climate swings do not explain everything bad that happens, and at the time, there seemed to be no particular reason to pose the question.

Indeed, it is when pathogens free themselves of the ecological constraints imposed by their animal hosts that they become most dangerous. Over thousands of years, for example, tuberculosis, influenza, measles, chicken pox, and pneumonia have all developed the ability to spread directly from human to human. Many of the viruses that cause hemorrhagic fevers, including Machupo, alarm scientists because it appears that they might be on a similar learning curve.

The 1971 outbreak of Bolivian hemorrhagic fever that occurred in the city of Cochabamba, for example, appeared to have started when a young nursing student fell ill. The student had just returned from visiting her family in the Beni, and it was assumed that she had come into contact with infected rodents there. Before she died, she passed the disease on to four other people, three of whom also died. A young Bolivian pathologist at Seton Hospital became an additional casualty. He became infected when he stabbed his finger during the autopsy of one of the victims. A short time later, he too died. It was an awful death, C.J. told Terry. As his blood vessels developed tiny sievelike holes, the young doctor bled out in myriad places.

Like ghost stories around a campfire, the stories C.J. told had an impact on Terry as night after night he set out his traps to capture live rodents for testing. The objective was to determine the prevalence of Machupo infection in the resident rodent population, and to do that, he and C.J. needed to obtain blood samples from each and every animal. The problem was, it was sweltering hot in the Beni, and even though the scientists had set up a proper field laboratory and had masks and gloves with them, no one felt like putting on additional layers. At the same time, no one wanted to handle the potentially dangerous *Calomys callosus* without donning protective clothing.

So Terry and C.J. devised a plan. They figured they could get away with processing all the other rodents without taking elaborate precautions. The trick would be to figure out in advance which rodent was which. On the team was a Bolivian technician who was said to be expert at separating the dangerous *Calomys* from its many look-alikes. The technician was also known to have been exposed to Machupo virus before; he tested strongly positive for the antibodies that conferred immunity.

The technician tried his best to sort the trapped animals into two groups, *Calomys* and non-*Calomys,* but he made a mistake. A short time later, Terry casually opened a trap and saw a *Calomys* staring him in the eye. "He slammed the cage closed," C.J. remembered with a chuckle, "and he came running around the corner. You know, Terry's hair is not too dense in front, and it was standing up all around the edges. He was yelling 'That son of a bitch No Mas Rosas.'" Tomás Rosas was the name of the technician, and Terry wanted no more (*no mas*) help from him.

What to do? Proceed, the two scientists finally decided, but very, very carefully. This time they put on their protective gear, and a good thing, too. The *Calomys,* they discovered, had a huge spleen that brimmed with Machupo virus. "You didn't even have to test it to know it was infected," said C.J.

Later, as he and C.J. were parting company in the Miami airport, Terry asked C.J., casually, he thought, "What are the symptoms of Bolivian hemorrhagic fever?" Not to worry, C.J. reassured him, "unless in a few days you should happen to develop a really high fever of about 104 degrees Fahrenheit."

Shortly after returning to his home near Albuquerque, Terry started feeling ill. He took his temperature. It was 104 degrees. "Look in my briefcase," he told his wife, Nancy, "find C. J. Peters's card, and tell him I'm running a 104 degree temperature." As it turned out, Terry had not been infected by Machupo. He had picked up an unpleasant but treatable bacterial infection from drinking contaminated water. Soon enough he recovered and returned to Bolivia to continue with his normal fieldwork routine.

Toward the end of June, Terry recalled, C.J. tracked him down at one of his rare stops in a hotel with a telephone. "Terry, what are you doing back in Bolivia," C.J. exclaimed. "Don't you know there's a big disease outbreak in your own backyard?"

The disease C.J. was talking about, hantavirus pulmonary syndrome, was not caused by the same type of virus as Machupo, but scientists would soon discover a similarity that made Terry's spine tingle. The species of hantavirus responsible for the disease outbreak in New Mexico was carried primarily by the deer mouse, a distant relative of *Calomys callosus.*

After the outbreak in New Mexico subsided, Terry started wondering whether other types of rodents might harbor hantavirus as well. He started wondering what might turn up in the rodent collection housed by the Museum of Southwestern Biology.

In addition to being chairman of the biology department at the University of New Mexico, Terry was also head of the museum's mammal and frozen-tissue collection, which was tucked away in cramped basement quarters downstairs from his office. Superficially, the museum looked like a taxidermy display. The larger specimens, including a cougar, paced the tops of steel cabinets, while the stuffed pelts of mice and rabbits and weasels had been filed away in the drawers below. The museum also maintained a frozen collection of tissues and organs from animals Terry and others had collected throughout the world.

After the 1993 hantavirus outbreak subsided, a team of scientists at the University of New Mexico used some of this preserved material to establish that harvest mice from Mexico and Costa Rica carried antibodies to a completely different strain of hantavirus. The search widened. Soon other New World hantavirus strains were turning up everywhere, in South America and North America alike. The rodents that had crossed over the Bering Strait had obviously brought the virus with them, and as new species of rodents arose, so did new species of hantavirus.

WHAT TRANSFORMED hantavirus into a kind of symbol for the threat posed by emergent diseases was not the evolutionary drama that called it into being but the climatological drama that transformed a seemingly innocuous infection of deer mice into a lethal infection of humans. For unlike the Machupo outbreak in San Joaquin three decades before, the hantavirus outbreak in the American Southwest left a trail of meteorological bread crumbs that scientists were able to track all the way from New Mexico to the tropical Pacific.

The disease first showed up in the Four Corners region, so called because it is where the borders of New Mexico, Arizona, Colorado, and Utah meet. The first to sicken were two young Navajos, Florena Woody, a pretty, willowy woman of twenty-one, and Merrill Bahe, nineteen, the top runner on the Santa Fe Indian School track team. They were in love. They had a five-month-old son. Then the future they'd planned together slid out from under them.

In late April 1993, Florena Woody developed aches and pains that seemed to point to a bad case of flu. She burned with fever.

She drew her breath in gasps. Her mother took her to the local Indian Health Service hospital. An X ray revealed that her lungs had filled with fluid. A short time later, she died. By this point, Merrill Bahe was also feeling ill. He rallied, though, and drove with a cousin to Gallup, New Mexico, to attend Florena Woody's funeral. En route, he became desperately ill. A short time later, he, too, was dead.

What terrible agent could have mowed down two people who'd been so young, so healthy, so strong? What uncaring force could have callously snuffed out not one but two promising lives? These questions arose as soon as an official from the New Mexico Office of the Medical Investigator, or OMI as it's called, talked to Bahe's relatives, at which point big wheels started to turn. Upon learning that Florena Woody was about to be buried, the OMI investigator rushed to the funeral home, met with her family and asked for permission to autopsy her body. It was an unusual request, and in the context of traditional Navajo beliefs, even sacrilegious. But the circumstances were also unusual. The family assented.

By late May, as other cases of the mystery disease were beginning to surface, C.J. and his team joined the hunt for the killer, methodically running down a list of the most likely suspects. Early on they ruled out bubonic plague, Legionnaires' disease, and influenza, and soon they had to admit they were stumped. "It seems as if there's a hole in the sky through which bad things are pouring," C.J. recalled a Navajo medicine man saying. There was a kind of hole in the sky, of course; it was the hole through which rain was pouring. But it took a while for this to be understood.

What was frightening about the disease was its rapid progression, from aches and coughs, nausea and fever, to death in almost no time at all. Among the symptoms were extremely high white-blood-cell counts and lungs so congested with fluid that victims, in effect, drowned. The two were undoubtedly related. For the body does not allow invading microbes to go unchallenged; white blood cells, armed with potent enzymes, engage these, humankind's most ancient foes, in sometimes all-too-mortal combat.

The chemical warfare the immune system conducts works most of the time, but occasionally it backfires. Some of the compounds white blood cells unleash make capillaries leaky, enabling plasma to seep out of the bloodstream and into vital organs like the lungs.

Any infectious agent that could provoke such a potent reaction was clearly dangerous. Working out of a biohazard level 3 lab in Atlanta, the special-pathogens team took the precaution of donning plastic coveralls, respirators, and plastic visors before processing samples of blood and tissue taken from those who had died. One of the first things they did with the samples was mix them with antibodies that homed in on different viruses, among them Machupo and Lassa, which belong to the arenavirus genus, as well as hantaviruses like Hantaan and Seoul. The antibodies were fluorescent, so that whenever they latched onto a virus, the samples would glow under a UV lamp. None of the arenavirus antibodies stuck to the samples. But the hantavirus antibodies did, and when they did, they glimmered like microscopic neon signs.

Hantaviruses, C.J. knew, had plagued U.S. troops during the Korean War—Hantaan is named for a river that runs through the DMZ—and they were all carried by rodents. The Seoul virus had, in fact, been found in Norwegian rats trapped in the United States, though it had not been previously associated with outbreaks of human disease in North America. The Centers for Disease Control investigator who'd combed through Merrill Bahe's deserted trailer, C.J. also knew, had found a peppering of rodent droppings along the floor. Still, that in itself was not surprising: in rural areas, when people move out of dwelling spaces, rodents aggressively move in. The mere presence of rodents did not clinch the case.

Plus there were pieces of the puzzle that didn't fit. Hantaan and Seoul attacked primarily the kidneys. Severe cases, marked by hemorrhaging, could be fatal, but 80–90 percent of those infected eventually recovered. The virus that first advertised its presence in the Four Corners, however, targeted the lungs. Barely 50 percent of its victims survived. At that point, C.J. began to wonder if his team might have discovered a new, previously unknown strain of hantavirus. It was just a hunch, but it turned out to be right. Once his team had decoded the virus's genome, it became quite clear that the exact sequence of "letters" that composed the genetic script had never been seen before. The mystery killer was, very clearly, a New World variation on an Old World theme.

According to convention, the virus should have been named after the place where the first cases arose, but residents of the area, concerned about the potential impact on tourism, objected to the idea of calling it the Four Corners virus. C.J. then proposed Muerto Canyon as a substitute, but that name was deemed to be too close to a landmark in the Canyon de Chelly National Monument. Appropriately enough, the virus no one wanted to claim came to be known as Sin Nombre, the virus with no name.

Like all viruses, Sin Nombre was a curious entity, not exactly living but also not exactly dead. Structurally, C.J. noted, it consisted of little more than a coil of RNA, a close cousin of DNA, packed, as all viruses are, into a distinctively shaped capsule. As soon as it infected a cell, however, this deceptively inert bundle of genetic material sprang into action, seizing control of cellular machinery to copy itself a billionfold. All viruses are like that, obligate parasites that cannot maintain themselves in the absence of an animal host. So that raised a question: What animal or animals had spread Sin Nombre to humans?

Barely a month after Florena Woody and Merrill Bahe died, teams of biologists fanned out across the Four Corners region and, under the unrelenting sun, started trapping mice, rats, prairie dogs, rabbits, even skunks. In short order, they drew blood samples from the trapped animals and shipped them off for analysis. In this fashion, the chief carrier—*Peromyscus maniculatus,* the deer mouse—was very quickly identified. Indeed, 30 percent of the deer mice the field teams caught were infected with the virus. Furthermore, the virus carried by the mice turned out to be genetically identical to the virus that had been making people so sick. And whenever field workers set up traps in the homes of those infected by the virus, the animals that ended up in the traps were overwhelmingly deer mice.

From the point of view of the virus, these pert little animals made excellent hosts. They were, first and foremost, extremely prolific. Females bore multiple litters, sometimes as many as three or four a year, and while the first litter a female bore was typically small, sometimes only a single pup, the size of each successive litter increased. Gestation was a streamlined process that took only twenty-one to twenty-eight days, and the pups, naked and blind at birth, grew so quickly that they could be weaned at three weeks.

In addition, deer mice were opportunistic omnivores. They could, and did, dine off not just seeds and nuts but also crickets, beetles, and other insects. They were known to inhabit dryland habitats across the United States, save for the far Southeast and a thin strip along the Atlantic seaboard. Their very ubiquity increased the chances of close encounters with humans, particularly those who lived in rural areas. Indeed, the majority of people sickened by hantavirus in 1993 lived in areas with no more than five homes within a half-mile radius.

Like the Machupo virus, hantavirus appeared to have no need of fleas or mosquitoes or other insects to assist in its spread. It traveled through the air, in the form of aerosolized specks of mouse urine and feces, so that people became infected simply by breathing. People unwittingly risked exposure whenever they entered structures inhabited by infected mice; their risk of exposure increased in close places where air was unable to circulate freely—in crawl spaces under houses, for example, or in the sealed-up rooms of infrequently used cabins.

Hantavirus pulmonary syndrome was, fortunately, a very hard disease to catch. Unfortunately, it was life threatening when one caught it. In the early stages of infection, giving patients antiviral medication appeared to be somewhat useful. But in the later stages, there was nothing to do but put patients on heart-lung machines that oxygenated their blood, and hope for the best. Either their bodies would fight off the infection and their lungs would be able to function again or they wouldn't. Despite heroic efforts to save them, about half still died.

The tragic deaths of Florena Woody and Merrill Bahe focused attention on a heretofore-unappreciated health threat that existed nationwide. By the end of 1993, fourteen states had reported 53 cases of hantavirus pulmonary syndrome; by early 1999, the count had risen to forty states and 211 cases. Among those who died were a graduate student from California who contracted the disease while staying in a cabin in the Sierra Mountains and a man from New Jersey who became ill, experts speculated, while cleaning out a vacant warehouse.

In 1993, however, New Mexico, Arizona, and Colorado did have more cases than all the other states put together—thirty-four in all—and that raised an interesting question. Was the clustering in space and time just a statistical fluke, or was there a reason for the pattern?

As LUCK WOULD have it, a team of University of New Mexico biologists, led by Terry Yates and his colleague Bob Parmenter, had already collected the data that pointed to an answer. Since 1989, for purely scientific reasons, they had been keeping track of rodent densities at the Sevilleta National Wildlife Refuge, near Albuquerque. Twice a year—once in May and June, again in September and October—they set up trapping webs in the exact same locations, the point being to cover the full range of habitat, from piñon-juniper woodland and mixed-grass prairie to Chihuahuan Desert scrub. At the same time, they collected detailed measurements of precipitation from weather stations located near each trapping zone.

It was all part of a long-term project aimed at understanding the dynamics of ecological change, explained Bob Parmenter, head of the Sevilleta field station, and in 1993, those dynamics were driven by El Niño. As it happened, the rodent study started during 1988–89, a strong La Niña year. From October 1988 to May 1989, the Sevilleta recorded a scant 37 millimeters of rain—

not even 2 inches. In 1991–92, by contrast, it recorded an astounding 237 millimeters during the same time period, more than double the mean of 108 millimeters.

El Niño, of course, was the reason that New Mexico enjoyed an exceptionally wet winter in 1991–92, followed by an exceptionally wet spring. The additional moisture, supplemented by a robust summer monsoon, caused the alfalfa in farmers' fields to grow tall and lush; it also created a cornucopia of wild seeds, berries, acorns, and nuts. Strengthened by the additional food, adult rats and mice produced extra litters. Then, in 1993, courtesy of the continuing heat wave in the Pacific, the mice enjoyed mild, moist weather through late winter and early spring. The moisture in 1993 was helpful, too, because a lot of it fell as snow, which formed an insulating layer over the nests the mice had built in the ground. As a result, more mice than usual survived into spring.

The scientists' conclusion—that climate and not some other factor was the most important variable—gained additional credibility from a similar project that had been conducted in Canyonlands National Park, near Moab, Utah. Since 1987, biologists there had been trapping mice in grasslands and woodlands that were virtually identical to those found in the Four Corners region. In contrast to northern New Mexico, however, Canyonlands had experienced atypically dry weather in 1991 and 1992. As a result, the scientists found, in three out of four of their Canyonlands sites, the mice and rats were barely able to hold their own, and in some cases their numbers even decreased. In 1993, New Mexico reported eighteen hantavirus cases; Utah, not a single one.

In the wake of the 1993 hantavirus outbreak, the University of New Mexico helped set up a trapping system around the state to monitor the status of the state's rodent population. For the next few years, the numbers of rodents remained low, and so did the number of hantavirus cases. Then, between March 1997 and March 1998, the population of rodents at some of the trapping sites increased ten- to twentyfold. The percentage of infected animals was also starting to angle back up.

Spring was always the time of peak danger. It was the time when the shelves in the wild larder were emptiest, having been substantially depleted during the long months of winter. It was thus the time when mice and rats were hungriest and, in their search for food, most likely to come into contact with humans. At the end of April 1998, well after El Niño had climaxed, hantavirus claimed the life of a twenty-three-year-old construction worker from Nageezi, New Mexico. His name was Hardy Haceesa. He was a devoted husband and father. He was the star center on the Nikes basketball team. Sadly, he was the first person in New Mexico to die of hantavirus pulmonary syndrome in more than four years.

DÉJÀ VU?

"Hurry, Nancy. Hurry," Terry Yates sang out to his wife. "You're only allotted so many sunsets in life, and it's a shame to miss one."

It was early July 1998. Terry was standing with his back to his house, a glass of bourbon in his hand. In the distance he could see Cabezon Peak, a rounded basaltic spire sacred to the Navajo. A cloak of clouds partly obscured the remnant throat of the ancient volcano, and as the sun dropped out of sight, it backlit streaks of rain that shimmered like tinsel suspended in midair. Virga, such rain was called, after the Latin word for twig or streak. It was a capricious phenomenon; the droplets that constitute a virga shower were weighty enough to fall from the clouds but too small to resist evaporation. In the hot, dry air of the desert Southwest, it was not at all unusual for rain to pass directly overhead without ever moistening the ground.

In July 1997, Terry remembered, the rains that rode the wings of the summer monsoon had been late, which struck everyone as odd, given the power of the new El Niño that had just taken off like a rocket. Day after day, temperatures soared above the 100 degree Fahrenheit mark, and by eight in the morning, the sun blasted out heat like the broiler in an oven. Toward the end of the month, though, the rains came. Virtually overnight, one of the driest Julys became the wettest July some places had experienced in almost ten years.

And that had been just the beginning. In fall the remnants of powerful Pacific Ocean hurricanes sallied much farther inland than usual, dumping even more rain on the desert Southwest. Then, after the hurricanes, came the extratropical cyclones that made landfall in California and clipped New Mexico as they wended their way cross-country. Granted, the 1998 winter storm season had drawn to a close earlier than expected, at the end of March as opposed to April or May, but by then the rodent population was already on the rise.

Now midway through the summer of 1998, the rodents were entering a time of plenty. Out on the Sevilleta National Wildlife Refuge, clumps of rice grass bent under the weight of silvery seed plumes, and nightshade spilled seedpods, puffy as popcorn, all over the ground. A lush stand of needle-and-thread grass was just drying out; its seeds, still tinged with green, would soon be ripe. Plus in the higher elevations, piñon pines were masting, swelling with nuts, that is. "We're going to have a bunch," Terry said. "Piñon pines don't make a crop every year, you know. Rainfall has some effect, but the real kicker is whether it gets real cold in October two years before they make their crop. Then they have to have enough moisture after that."

In 1993, Navajo medicine men had in fact alluded to outbreaks of disease that had occurred in the past, during times when pine nuts were noticeably abundant. But as Terry saw it, the pine nuts were just an additional goody for the rodents. "Their increase is driven by seeds from the grasses, which respond much more rapidly to climate change. El Niño causes more moisture to come to this part of the world in late winter and early spring, and these grasses respond by being fruitful and multiplying." And so do the rodents that eat the seeds.

That there was a causal chain connecting the El Niño cycle to the 1993 hantavirus outbreak seemed clear, but scientists had yet to work out the links. Some, for example, thought that the severe southwestern drought associated with the 1988–89 La Niña might have set the stage by drastically reducing the rodent population, thereby sending rodent-predators like coyotes and owls into a tailspin of their own. Predators, which typically bear fewer young and take longer to raise them than the species they prey upon, would have taken longer to replenish their numbers once the rain returned, or so the reasoning went.

There were other biological factors involved as well. Terry, for example, often wondered about the role turf battles between males might have played in priming the pump for the 1993 outbreak. By itself, he noted, an increase in the deer mouse population was insufficient to power an outbreak of disease among humans. Before that could happen, there had to be an increase in the number of *infected* deer mice, and that raised a question. Unlike the floodwater *Aedes* that carry Rift Valley fever, infected female deer mice did not appear to pass hantavirus onto their pups either in utero or during the process of birth. How, then, did adult deer mice become infected?

What Terry found intriguing was a smattering of observations that suggested older male mice were far more likely to carry the virus than younger male mice and females. The more he thought about it, the more likely it seemed that the 1993 outbreak might have something to do with an uptick in the number of fights between male mice competing for turf.

The scenario went like this: Infected older males transmitted the virus to younger males in the course of biting and scratching. These newly infected males, in turn, passed the infection on to the females. Then spring, for rodents the cruelest season, arrived. Eventually, Terry speculated, food stress might have built up to the point that it triggered a phenomenon known as dispersal.

The best-known example of dispersal occurs among rodents called lemmings, usually after a population increase outruns the resources of the local environment. In response, the lemmings start to move, at first stealthily, under the cover of night. But as a kind of hysteria takes hold, they abandon all caution and rush forward in frenzied mobs. In Norway, lemmings have been known to drown as they plunge over cliffs and into the sea, pushed to their deaths by the animals behind them. Deer mice didn't surge across the New Mexican landscape so dramatically, Terry thought, but they did wander much farther afield than one might imagine, sometimes covering a good half-mile in a single night. If on their journeys these mice happened across a house or a cabin or a trailer home, they were very likely to squirm their way in.

The problem was, biology is complicated. There were a maddening number of variables that might have been involved in the 1993 hantavirus outbreak; the trick lay in determining which were consequential, which coincidental. That was where the stainless-steel enclosure Terry jokingly referred to as the VLMA, or very large mouse array, came in. It was a very New Mexican play on words. The VLA, or very large array, is a network of radio telescopes aimed at the cosmos, and it lay just beyond the border of the Sevilleta National Wildlife Refuge.

The VLMA was the size of thirty-two football fields, and it wound up and down the rolling contours of a sandy rock-and-brush-strewn hill. Wire-mesh fencing, reinforced with solid sheet metal, delineated four identical quadrants, each home to a flourishing population of rodents. These ranged from desert specialists like pocket mice and kangaroo rats to the hardy opportunists of the *Peromyscus* genus: the deer mouse, the white-footed mouse, the brush mouse, the piñon mouse. (The *Peromyscus* genus belongs to the family Muridae.) There were plenty of wood rats around as well. Their twiggy nests formed messy piles on the ground.

The array was designed to keep rodents in, predators out. Above the primary fence, a series of electrified wires thwarted midnight raids on the rodent pantry by coyotes, bobcats, gray foxes, and ringtail cats. As for the snakes that used to live inside the array, those were being unceremoniously removed one by one. "I think we've pulled out a couple of rattlers so far," said Terry.

"One was a diamondback over five feet long. The idea is that the array should be predator-free except for hawks and owls, and since mice and rats don't tend to be up and running around much in the daytime, hawks really aren't a big problem."

In part, the removal of predators was a convenience; it ensured that Terry and colleagues would not run out of captive subjects to study. But it also afforded a way of testing the impact a scarcity of predators might have on the resident population of rodents.

In July 1998, the array had been up and running for about a year. At the time, field teams were in the midst of conducting a midyear census of all the rodents that lived within its confines. The night before, the field teams had set out galvanized steel traps, and now they were rushing to collect them, before they turned into the equivalent of solar-powered ovens. The first order of business involved coaxing each trapped rodent out of its cage and into a clear plastic bag. The second involved seizing the rodent by the scruff of its neck, then measuring and weighing it. The procedure didn't hurt, but the mice often squeaked in alarm as they dangled by one foot from calipers attached to a portable scale.

Soon afterward, the captive rodents were loaded onto the back of a pickup truck and rushed over to the Sevilleta's laboratory. There technicians anesthetized the animals and drew blood and tissue samples. At the end of the day the field teams then carted the mice and rats back to the VLMA and released them near the places where they had been caught.

Already it was clear to Terry that the situation in New Mexico in 1998 was not exactly analogous to 1993. For one thing, while rodent densities were high, they were not as high as they had been in 1993. For another thing, between 1990 and 1993, an unusually extended El Niño had brought the Four Corners region copious rains two years in a row. By contrast, the 1997–98 event was short-lived; in July 1998, in fact, the tropical Pacific was beginning to flip to cool La Niña conditions. Moreover, while the rains in 1997–98 had been generous, they did not reach the extraordinary levels recorded in 1991–92, when the Sevilleta received more than twice its normal ration of rain.

Thus far just three people in New Mexico had contracted hantavirus, and Terry, for one, was beginning to think that the big risk of an outbreak might lie ahead, in the springtime of 1999, as El Niño's opposite, La Niña, tightened the noose of drought around the region. A prolonged drought, he figured, would trigger a truly desperate search for food, sending mice scurrying into dwellings. Would that really happen? Yes, declared Terry, "unless the kangaroo-rat population goes up and outcompetes the deer mice, unless La Niña causes the rains to start a month late so that the food resources are all used up and the mice die, unless the coyote population goes up." There were, in fact, ten cases of hantavirus pulmonary syndrome in New Mexico in 1999, five of which were fatal, compared to six in 1998, with four deaths. Though it seemed he was right, even Terry did not really know why.

THE BALANCE OF nature is dynamic, not static, so that each twist of the climatological kaleidoscope inevitably catches plant and animal populations in a slightly different state, and in the case of disease outbreaks, cultural variables can be even more important than their biological counterparts. Thus, while an uptick in the population of infected deer mice would seem to be necessary to power an outbreak of hantavirus, it is also not sufficient. Human beings, after all, have the ability to lower their risk of exposure by taking any number of precautionary measures, including mouse-proofing their dwellings and airing out closed spaces before entering.

Over the years, a variety of simple measures—mosquito nets, screened porches and windows, filters for drinking water—have consistently shown great power to damp down insect- and water-borne diseases like malaria and cholera. That these diseases continue to threaten human lives, many experts believe, has as much to do with social factors as with climatological ones. The Rift Valley fever epidemic in East Africa, for example, would not likely have taken so many human lives had people not been so ignorant and desperately poor that they were willing to eat animals that were clearly sick.

Both Terry and C.J. were intrigued by the multiple ways in which people so often—and so unwittingly—exposed themselves to harm. "Imagine a big forest and a strip of savanna," said C.J. "And then imagine that you put a little house down on that savanna, and then imagine that you build a bunch of houses. What happens is that there are a bunch of rodents living out there and some, though not all, will move in." Like North America's *Peromyscus maniculatus,* Bolivia's *Calomys callosus* liked to move into people's houses. To paraphrase the line from the movie *Field of Dreams,* if you build it, they—in this case, rodents—will come.

People controlled deadly diseases when they curtailed their animal carriers' ability to find niches in which to breed and thrive; they allowed deadly diseases to explode when they afforded them opportunities to expand in numbers. People planted food for rodents in their gardens and fields; they stored food for rodents in their kitchens and granaries. They also provided habitat for mosquitoes and warm-blooded animals, including themselves, for mosquitoes to bite. And last but not least, people also gave viruses and bacteria and the animals that carried them free lifts around the world.

That would become apparent three years later, in the year 2000, when an outbreak of Rift Valley fever unexpectedly hit Saudi Arabia. Probably, C.J. thought, some of the mosquitoes infected with Rift Valley fever had radiated out from the epidemic in East Africa, perhaps by hitching a ride on a boat or a bus filled with people and animals. Like the big block of unstable rock hanging over the town of Rio Nido, the microbes unleashed by the last El Niño of the twentieth century continued to haunt the twenty-first-century world.

CHAPTER **EIGHT**

All men live enveloped in whale-lines. All men are born with
halters around their necks; but it is only when caught in the
swift, sudden turn of death, that mortals realize the silent,
subtle, ever-present perils of life.

HERMAN MELVILLE, Moby Dick

WANARISET SAMBOJA

Adi Susilo let his fingers slide along the silvery buttress of a *Shorea laevis,* a skyscraper of a tree with lichen-marbled bark. "We still have something," he wistfully said. A slender, bespectacled man in his late thirties, Adi was the official in charge of Wanariset Samboja, a forestry research station that lies along a busy tarmac road in the Indonesian province of East Kalimantan. Adi had grown up across the Java Sea in the former royal city of Solo, and the trees he was charged with protecting reminded him somehow of the princesses who dreamily danced to the gamelan music that filled the pavilions of Solo's palace. To Adi, trees like the *Shorea laevis* were, above all else, elegant.

It was October 1998. The fires that had blazed across East Kalimantan's rain forest were now a bitter memory, as was the El Niño that had set the stage for the fires by enveloping the area in drought. After an exceedingly short interregnum, a rambunctious La Niña had succeeded in shoving her more famous sibling aside, ushering in the climatological equivalent of a counterrevolution. To those places that El Niño had made wet, clear skies returned, while in East Kalimantan rain fell in almost inconceivable deluges. In August, in fact, some one hundred thousand people had to be evacuated from their homes in the outskirts of the city of Samarinda after an overstressed dam burst, putting thirty thousand houses under as much as 8 feet of water. Roads now resembled muddy quagmires; a tangle of lush new growth concealed fire-charred ground; the Mahakam River had surged back to life again.

East Kalimantan is one of four provinces that make up greater Kalimantan, as the Indonesian sector of Borneo is known, and the Mahakam River is the major commercial artery that links the rugged interior of the province with the urbanized coast. The river's source lies very close to the equator, in the mountains of the densely forested Müller range, and from there it courses some 400 miles downhill, to Samarinda, a booming timber center that may well boast the highest density of sawmills in the world. Beyond Samarinda, the Mahakam fans out into a swampy delta that empties into the Strait of Makassar.

The banks of the lower Mahakam are lined with villages whose watery strip malls boast floating structures that cater to every basic need. There are floating restaurants, floating fuel stations, floating grocery stores; there are even rows of floating shacks that serve as communal latrines. Along this bustling thoroughfare, ferries and small, swift *longbots* ply routes up and down, carrying passengers and freight of astonishing variety, from bananas and melons to satellite dishes and motorbikes. Like sports cars, the *longbots* deftly weave in and out of an endless procession of slow-moving barges heaped high with coal and rafts of logs so massive that people actually live on them, sleeping under tarmacs for shelter.

In 1997, however, as drought settled upon the Indonesian archipelago, traffic on the Mahakam dramatically slowed; in places, it even stopped. The trouble started in May, when the dry season came in even drier than usual. Then the rainy season that normally commenced in October turned dry as well. Gradually, the Mahakam receded from its banks. Its oxbow lakes turned into foul-smelling marshes choked with weeds. The churning rapids that lay between the villages of Long Bagun and Long Pahangai died to a wispy trickle. For months not even the most daring *long-bot* operators attempted to negotiate the treacherous passage.

Most terrible of all was the punishment the drought meted out to the forests within the river's wide drainage basin. Near the town of Melak, the trees in an internationally acclaimed orchid preserve thirsted for rain, as did the trees of Kutai National Park and Wanariset Samboja. Their supposedly evergreen leaves turned autumnal shades of orange and yellow, the color of the flames that would soon come to consume them, and as they fell to the ground, they made a dry, crackling sound. Chiitt. Chiitt. Chiitt.

The fires that were to scourge the Indonesian archipelago broke out first on the island of Sumatra, shooting past the supposedly protected perimeters of Gunung Leuser National Park. Then Kalimantan burst into flames. By September 1997, fires were blazing out of control in the forests and peat bogs and coal seams of Central and West Kalimantan. In October, orbiting satellites started picking up suspicious hot spots in East Kalimantan as well.

Collectively, these fires cast a pall that extended over an area of 400,000 square miles, affecting not just Indonesia but Malaysia, Singapore, and parts of Thailand as well. The ash they rained down turned rivers acidic, so that fish broke out in sores and rashes. The smoky haze that spewed forth caused respiratory distress among an estimated 70 million people and compromised visibility to such an extent that a commercial jetliner crashed into a hillside near the Sumatran city of Medan and a ferry rammed into a freighter in the narrow seaway that runs between Malaysia and Sumatra.

And then, for a tantalizing moment, the danger seemed to ease. In November and December, just enough rain fell to damp the fires down. But the relief proved to be illusory, and soon thereafter the fist of drought clenched shut. Virtually always, Adi recalled, even at the height of the dry season, the forest at Wanariset Samboja retained enough residual moisture to drive the process of evaporative cooling. Yet now, in the middle of the wet season, this reliable natural air conditioner had inexplicably faltered. "For the first time I can remember," said Adi, "when I walked into the forest, I felt hot instead of cool."

Leaf litter, dry as parchment, piled up on the floor of the forest. A large peat bog turned into a shriveled-up sponge. And most unnerving of all, bloodsucking leeches, those ubiquitous pests, vanished from the forest floor. It was strange to stroll along a path and not encounter even one of these needlelike creatures trying to wriggle its way through the weave of a sock or the eyelet of a shoe. It was stranger still to find oneself concerned about what had happened to these loathsome pests and praying for their return.

In January 1998, Adi and his staff played a hard game of tag with a couple of small fires that raced in from the perimeters of the reserve. They used chain saws to cut firebreaks; the fires stopped at the breaks. But then smoke loomed up from behind a small hill deep within the interior of the reserve, and by the time they saw it, it was already too late. "We fought that fire for three days, and still there was fire," said Adi. "We couldn't even tell where the fire began or where it ended."

It was the kind of fire that all foresters dread, the kind that not only races along the ground but also flares across the canopy. Such fires (they are called crown fires) can jump 7 miles in less than half an hour, leapfrogging from ridge to ridge, launching flaming projectiles that seed new fires. All that remained for the situation to career out of control was for the winds to pick up, which they did. Soon flames flickered in the peat bog; they tap-danced on the tops of the trees; they burst from hollowed-out trees like colorful fountains. "It was very beautiful," Adi recalled, "and also very dangerous."

The flames generated tornadic vortices that sucked in air from all around; they generated noises reminiscent of a jet engine's roar, a freight train's rumble; they generated clouds as moisture boiled out of leaves and branches and steamed into the air. Once, Adi was trapped by smoke so thick that he lost his way; as he raced to the top of a hill, he could hear the fire panting behind him. He had been extremely lucky, he later reflected. Fires often leap up-hill with frightening speed because their rising heat dries out the vegetation forward of the flames. For that reason, steep terrain sharply increases the likelihood of a blowup, an explosive phe-nomenon that occurs as temperatures approach the point of spon-taneous combustion.

January, then February, passed without so much as a drop of rain. Physically and psychologically, Adi was nearing exhaustion. Sometime in March he called a meeting in the field tent that served as the fire command post. There he announced that he be-lieved that further effort was futile, that it would be best to give up and let the fires burn. His voice choked. His eyes brimmed with tears. Then members of his staff spoke up one by one. Let us make one last stand against the fire, they said.

At that point, Adi and his colleagues changed their strategy; in-stead of trying to be everywhere at once, they concentrated their efforts on saving a single 120-acre plot. Armed with not much more than chain saws and hand tools, a hundred people managed to beat back the fires that threatened a particularly spectacular stand of trees, including the *Shorea laevis* against which Adi was now leaning. The *Shorea laevis* was a memorial of sorts, he ex-plained. It was a memorial to the other trees that he and the oth-ers had tried to save. It was a memorial to the war they had fought against fire and lost.

OUT OF A TOTAL of 27,000 acres that fell under Wanariset Sam-boja's protectorate, some 20,000 acres burned, and prominent on the list of casualties was a test plot that was regarded as a show-case. It represented years of work on a problem that had intrigued Adi ever since he spent a year at the University of Florida in Gainesville, studying for his master's degree. For his thesis, he chose to examine the ability of tropical forests to regenerate after fires, specifically those that raged in nearby Kutai National Park during the 1982–83 El Niño, when rainfall dropped to a third of the annual mean. Eventually, young trees did spring up in the burned-over areas, Adi found, but the process of regrowth was excruciatingly slow, for the fires had done more than kill many of the mature trees in the forest. They had also completely sterilized the top layer of soil.

This layer of soil, Adi explained, was like a bank vault that contained a vast store of wealth. Only in this case the wealth was biological. It included the seeds that were about to sprout into the next generation of trees; the fungi that decompose wood, releasing chemical nutrients; the root-dwelling microorganisms known as mycorrhizae that supply trees with essential nutrients such as nitrogen. A forest that had lost its top layer of soil, in Adi's way of thinking, resembled a country with an exhausted treasury that was trying to spring back after a long, vicious war.

Around 1990, the foresters of Wanariset Samboja began experimenting with ways to spur the regrowth of trees on fire-scorched ground. For a while they tried gathering seeds from fruiting trees in the forest and planting them in pots, beneath wire screens that mimicked the canopy's filtered shade. But the seeds were finicky. Many of the seeds did not germinate at all, or if they did, the seedlings they produced proved fragile and failed to thrive.

The foresters then devised a different approach. They began going into the forest and scraping up soil, filled, they hoped, with the right kind of growth-stimulating microorganisms. Into pots filled with forest soil, they transplanted what they called "wildlings," little saplings that they had dug up from the forest floor. By 1997 the test plot at Wanariset Samboja boasted a number of wildlings that had grown into trees 20 or more feet tall, with luxuriant crowns, as dense as umbrellas, filled with glossy green leaves. Then, in 1998, the fires came and killed the wildlings, every one, leaving their lifeless trunks and leafless

branches standing like tombstones in a botanical Flanders
Field.

IF ONE LOOKED at all closely, one saw that Wanariset Samboja was
strewn with memorials to the recent war against fire, and among
the most poignant were the char marks high on the concrete wall
that surrounded the main cluster of buildings. Whipped by high
winds, the fires had raced in, full of fury; the wall had stopped
them just short of a complex of cages that sheltered dozens of or-
phaned orangutans. The young primates were there because, in
addition to the conservation of the surrounding forest, Wanariset
Samboja had added to its mission the rescue and rehabilitation of
orangutans captured by poachers for sale to foreign buyers or
kept by local families as pets.

With their flamboyant orange hair and Picassoesque faces,
orangutans are spectacular creatures. They are also the world's
largest tree-dwelling mammals, feeding and sleeping 100 or more
feet above the forest floor. Orangutans are believed to have
branched off the ancestral primate line some 7–10 million years
ago, and along with chimpanzees and gorillas, they are humankind's closest living relatives. That kinship is acknowledged
in their name: In Malay, *orang* means person; *utan,* forest. In the
distant past, orangutans had been broadly distributed across
Southeast Asia, but for a very long time they have been geographically confined to the rain forests of Sumatra and Borneo.

These forests are known as dipterocarp forests because they are
dominated by a large genus of trees with two-winged seeds. (In
Latin, *di* is two, *ptero* is wing, and *carp* is seed.) In addition to
dipterocarps like *Shorea laevis,* they include many other types of
trees, including sturdy ironwoods capable of resisting rot for several hundred years and wild relatives of the cultivated trees that
produce mangoes and mangosteens, breadfruits and durians. The
diversity of these forests seems essential to the survival of the
orangutans, which have an inordinate appetite for fruit.

The nineteenth-century naturalist Alfred Russel Wallace, an

evolutionary biologist whom many rank alongside Darwin, spent hours observing orangutans as they swung through the interlocking branches of towering trees. The animals rarely descended to the ground, he noted, except when driven by thirst or hunger. "Now it seems to me probable," Wallace wrote, "that a wide extent of unbroken and equally lofty virgin forest is necessary to the comfortable existence of these animals. Such forests form their open country, where they can roam in every direction with as much facility as the Indian on the prairie or the Arab on the desert; passing from tree-top to tree-top without ever being obliged to descend to the ground."

Unfortunately for the orangutans, lumber and plywood from dipterocarp trees turned out to be a commercially valuable commodity, as did the gold, the coal, and the oil and natural gas that lay beneath the tree-covered surface of the island and the tepid waters of its shallow seas. The land that the trees occupied was valuable, too, as once it was cleared, it could then be converted to other uses. As a result, all over Kalimantan, coal mines, gold mines, oil wells, farm fields, and tree plantations were sprouting in place of the forests that the orangutans called home.

Everywhere, it seemed, once-dense stands of dipterocarps were being replaced by fast-growing eucalyptus and acacia and oil palms and rubber trees and by expanses of alang-alang, a coarse grass that very quickly invades fire-scorched land. With its thick subterranean mat of roots and rhizomes, alang-alang itself was virtually impervious to fire; indeed, it flourished in precisely those places where fires were frequent and trees scarce, and for good reason. Among the few things capable of killing the bright chartreuse spears of alang-alang was shade.

Prior to the 1997–98 El Niño, orangutans already were in trouble due to the dwindling of their habitat. But then, the extended drought and the fires that followed escalated what had been a chronic problem into a crisis. Emboldened by hunger or panicked by fear, adult orangutans—many of them females with babies clinging to their necks—ventured out of their forest homes and into villagers' gardens and fields. In the confrontations that frequently ensued, the adult animals were sometimes killed, their infants captured and put into cages.

In early 1997, Wanariset Samboja housed just a few dozen orangutans, a manageable number. Not quite eighteen months later, the number jumped to almost two hundred. Many of the animals were extremely young, the size of human toddlers, and they hugged the legs and swung from the arms of technicians who served as surrogate mothers. The technicians were all female, and they wore surgical masks to avoid infecting their charges with a cold or other infection. Orangutans, it turns out, are so similar to humans that they easily pick up human diseases, including influenza, hepatitis, and tuberculosis.

The objective at Wanariset Samboja was to keep the animals healthy until they were old enough and strong enough and also knew enough to be able to survive in the wild on their own. For this reason, the surrogate mothers tried to introduce their charges to foods that grew in the wild, peeling away the skin of a fruit or tender shoot to reveal the delicacy inside. By October 1998, in fact, more than forty young orangutans were judged to be ready to try life outside Wanariset Samboja's protective perimeters. By then, however, the fires had destroyed so much forest that it was hard to find places to put them.

At precisely this time a new problem materialized in the form of an adult female orangutan marooned in a patch of trees too small to support her. Hearing of her plight, a team from Wanariset Samboja rushed to the rescue. They brought along visiting veterinarian Henning Wiesner, who was not only the head of the Munich Zoo but also an expert in the use of tranquilizer guns. Wiesner had even concocted his own special slow-acting potion so that the animal did not fall from the tree to which she was clinging but slowly climbed down. She was about seven or eight years old, Wiesner figured, and so he named her Miriam, after his own eight-year-old daughter.

A day or so later, Wiesner and the others drove Miriam to her new home, a journey of about four hours over rough, rutted roads that passed through fields of alang-alang and timber concessions filled with fire-killed trees. By the time she arrived, Miriam was a study in rage. She grabbed the bars of her cage with strong, hairy fingers, bared her teeth, and unleashed a series of rasping grunts. As gently as they could, a group of men lifted her cage from the back of their pickup truck, carried it to the edge of the forest, and gingerly raised its gated door.

Miriam did not hesitate. She bolted out in an upright crouch, grabbed hold of the nearest tree, and clambered up it like a telephone lineman. Thirty feet off the ground, she stopped and turned around. Hooting in defiance, she broke off branches and threw them at her tormentors. Then, with acrobatic panache, she moved deep into the forest, swinging from treetop to treetop, a vibrant splash of orange in a field of green.

The forested fastness where Miriam was released, on the rugged slopes of the Meratus Mountains, was a place of breath-stopping beauty. Giant trees with flaring buttresses stood guard like sentinels; they wore garlands of lianas and corsages of orchids. Two hundred feet above the forest floor, the branches of these trees spread out to form a cathedral-like canopy, and their leaves filtered sunlight like stained-glass windows. This was one of the last large tracts of unbroken lowland rain forest in East Kalimantan; it was by definition, then, one of the largest tracts of prime orangutan habitat in the world.

Miriam didn't realize it, but she was extremely lucky. Across East Kalimantan, it was thought that a thousand adult orangutans had ended up in similar straits—and there weren't that many orangutans left. In 1900, it was estimated, there were some 315,000 orangutans in the world; by 1998, it appeared that their numbers had declined to less than 30,000. At most, it was thought, there might be between 5,000 and 7,000 orangutans left in Sumatra, another 3,000–5,000 in Malaysian Borneo, and maybe 15,000 in all of Indonesian Kalimantan. In 1997 and 1998, the combination of El Niño–triggered drought, deforestation, and fire edged the world's dwindling population of orangutans that much closer to extinction.

A TRAGEDY FORETOLD _____

An El Niño year straddles the usual January to December calendar. So while, in 1997, according to the World Wildlife Fund–Worldwide Fund for Nature, "more tropical forest burned around the world than at any other time in recorded history," it was really the El Niño year that began in the spring of 1997 and ended in the summer of 1998 that deserved to be immortalized as "the year the world caught fire."

During that time period, the organization noted in a report issued jointly with the IUCN–World Conservation Union, "a combination of the dry conditions caused by El Niño and uncontrolled burning practices took their toll on the world's forests." Fires exploded across Southeast Asia, from Malaysia and Thailand to Papua New Guinea and the Philippines. They raced across Australia, Africa, and South, Central, and North America. In Mexico and Nicaragua fires consumed some three million acres; in the United States, some two million acres. The hardest-hit state, ironically, was Florida, where one of the wettest winters on record gave way to one of the driest springs the residents had ever experienced.

The great majority of the fires that occurred in 1997 and 1998 were related in some way to El Niño–inspired droughts, including the fires that seared at least eight million acres of forested land in the Brazilian Amazon. Actually, the damage was probably far greater than that. Most of the fires that spread along the forest floor went undetected, but the damaging wounds they inflicted, researchers noted, would make it easier for the fires the next time to cut into the forest more deeply.

But of all the fires that sprang up in El Niño's wake, those that exploded across the Indonesian archipelago were in many ways the worst. Some twenty-three million acres in Java, Sulawesi, Irian Jaya, and Sumatra as well as Central, West, and East Kalimantan were stripped of vegetation by fire, and of that vegetation, approximately half was forest. The economic losses

Indonesia sustained were estimated to range from $5 billion to $10 billion. The biological losses were incalculable.

It wasn't that drought was an unexpected consequence of El Niño or the fires an unexpected consequence of drought. Virtually the same thing had happened fifteen years earlier, so that it was almost as if one were watching a reel of film that had been rewound and replayed. During the 1982–83 El Niño, fires of extraordinary intensity had also blasted through large areas of Borneo and Sumatra. Smoke from these fires had also poured out in roiling clouds that enveloped Singapore and Kuala Lumpur, and the visibility again was so poor that airports closed and ships were unable to venture out of port.

But the fires of 1997–98 were worse. They burned an area that was more than 50 percent larger. In East Kalimantan alone they destroyed 1,800,000 acres of selectively logged dipterocarp forest, 1,275,000 acres of tree plantation, bush, and degraded forest, 850,000 acres of peat-bog forest, 280,000 acres of primary dipterocarp forest, and some 300,000 acres of low scrub and grassland. The total affected area amounted to some 10 million acres, 15,000 square miles, an area almost the size of the Netherlands.

In Indonesia, the fires of 1997–98 grew to symbolize a corrupt social order that had careered out of control, and in the end, the choking plumes of smoke rising from burning forests combined with a drought-shriveled rice crop and the Asia-wide financial crisis to produce a storm of popular discontent, so that the upheaval in the natural world found uncanny reflection in the rioting that swept the streets of Jakarta. Caught in the crosshairs of the deepening malaise was President Suharto himself, who was forced to resign after losing his mandate to govern.

It was the Suharto government that set the stage for the fires when, around 1975, it started granting timber and mineral concessions in Kalimantan to large-scale commercial interests. The logging in the timber concessions was supposedly "selective" in that only certain trees were felled and the rest left standing, and in principle, the approach seemed to be an attractive, environmentally preferable alternative to clear-cutting. In practice, however, selective logging fell far short of its promise.

The problem was, when big trees crashed through the canopy, they often knocked down neighboring trees or sheared off their leaves and branches. Gaping holes appeared in the leafy sunshade created by the interlocking branches of extremely tall trees, so that moisture that previously had been trapped inside the forest was allowed to escape. At the same time, logging debris—branches and leaves stripped from the felled trees—built up on the forest floor. As the logging continued, the gaps in the canopy grew larger; woody shrubs, filled with flammable oils and resins, began to colonize the empty, sun-filled spaces.

But the structural changes in the landscape wrought by logging were only part of the story. There were other factors, among them government subsidies aimed at increasing Indonesia's production of pulpwood and edible oils. These subsidies actually made it financially advantageous for the owners of timber concessions to get rid of remnant forests (after extracting the largest, most commercially valuable trees) and to convert them into tree plantations filled with acacia, eucalyptus, and oil-producing palms. From their perspective, fire was just an easy, effective, and cheap way to clear logged-over land.

The pattern of fire in 1997 and 1998 strongly suggested that the subsidies played a role. According to the World Wildlife Fund–Worldwide Fund for Nature, some 80 percent of the fires that erupted in Indonesian Kalimantan originated on timber-concession and tree-plantation lands. The rest had causes that ranged from casually dropped cigarettes and imperfectly doused campfires to fires set by villagers to clear small plots for planting bananas, cassava, and maize.

THE FORESTS OF Kalimantan have never been totally innocent of fire. In 1989, German fire ecologist Johann Goldammer and his colleagues identified layers of charcoal left by fires that burned in East Kalimantan between 17,500 and 350 years ago. These fires, in other words, occurred not only at the end of the last glacial era, when climatological conditions are believed to have been much drier than today, but also within the current warm and wet regime—probably, Goldammer has speculated, when a lengthy El Niño–inspired drought randomly intersected with some source of ignition.

The source of ignition could have come from fires set by aboriginal hunters to flush out game or from slash-and-burn agricultural activity at the margins of the forest. There have been people living in this region of the world for quite a long time. The first inhabitants were a pygmy people who arrived around thirty thousand years ago, undoubtedly walking across the land bridge that once connected the island of Borneo to the Malaysian mainland. They were succeeded by successive waves of Austronesian migrants, who, around four thousand years ago, began arriving in sturdy outrigger canoes, probably from the vicinity of the Philippines.

There were also natural sources of ignition. In East Kalimantan, for example, Goldammer and colleague Berthold Seibert found a coal seam that had been set ablaze 15,300 years ago, by lightning, they presumed, and had continued burning for 2,000 more years. In case anyone doubted that a burning coal seam could ignite a fire, they provided an eyewitness account. During the El Niño of 1986–87, Goldammer and Seibert actually watched a burning coal seam set off a wildfire in a reserve not far from Wanariset Samboja. It had caught fire during the El Niño of 1982–83, they observed, and had been burning ever since.

But even though fires had erupted in the rain forests of Kalimantan before, in recent millennia, at least, they are thought to have been relatively few in number and modest in scale. They would have to have been; otherwise these forests would not have been dominated by fire-sensitive dipterocarps but by other types of trees better equipped to handle the assault of flames. Different trees have different tolerances for fire. Some—the ponderosa pines of North America, for example—actually depend on fires sweeping through often enough to get rid of the fuel load in the form of understory growth and dead trees killed by insects. Otherwise so much fuel builds up that a lightning strike or, for that matter, a cigarette tossed out of a car can touch off a fire of such high intensity that it kills everything in its path, including the healthy, mature trees that flourish when fires are more frequent.

But what is good for thick-barked ponderosa pines is not good for thin-barked dipterocarp trees, many of which cannot tolerate fires of modest or even low intensity. Dipterocarp forests, Goldammer has noted, are perfectly capable of growing in many parts of Southeast Asia, yet they are not widespread. He believes that frequent fires set by people have tipped the ecological advantage to fire-tolerant pines, so that forests dominated by dipterocarps have disappeared except in areas where fires have remained infrequent. In Kalimantan, the combination of a relatively sparse population of humans and a superabundance of precipitation helped damp down fire for a very long time.

In a desert environment, even a modest amount of humidity—20 percent or so—is enough to quench a grass fire, and the island of Borneo is normally far from a desert. Along the coast, which is the *least* rainy part of East Kalimantan, rainfall averages some 2,500–3,000 millimeters, or between 100 and 120 inches per year. Rain falls in the wet season, which runs from October through March, and it falls in the dry season as well. During the month of September, for example, the coastal oil city of Balikpapan might receive 170 millimeters (7 inches) of rain, which is a lot more rain than cities in the Peruvian desert usually see in the course of a year.

Dipterocarp trees, however, require a lot more moisture than trees adapted to drier climes, so that 170 millimeters a month is not as generous as it might seem. Indeed, if monthly precipitation falls below 100 millimeters, or about 4 inches, these trees soon become stressed enough to start dropping their leaves. Over time, leaf litter, flammable as paper, builds up on the forest floor, and desiccated branches and stems turn into kindling. Should a fire then break out, dipterocarps have few defenses. Not only are they thin-barked, but their woody tissues are also filled with flammable resins, and unlike many fire-tolerant trees, like euca-lyptus, for example, they lack the ability to resprout from the ground.

Dipterocarp trees are quite capable of beating back drought, however. Indeed, in the absence of fire, the sharp decrease in pre-cipitation occasioned by El Niños may actually serve as a catalyst for renewal. In a study conducted in West Kalimantan's Gunung Palung National Park, an international team of scientists recently uncovered a heretofore hidden connection between episodic droughts and a frenzy of reproductive activity among the resident dipterocarps. Exactly four times between 1986 and 1998, these scientists noted, the trees in their test plot produced large crops of fruit, and they did so in synchrony, seemingly cued in each in-stance by El Niño–triggered droughts.

Such sporadic mass productions of fruit are known as "mast-ing," and for countless millennia, masting would seem to have served the dipterocarps well. Its purpose, explained University of Michigan biologist Lisa Curran, is to ensure that the dipterocarps produce such an excess of single-seeded fruit that animals like orangutans and parakeets can never consume the whole crop. In-evitably, some of the fruits ripening in the trees are overlooked and allowed to produce viable seeds; inevitably, some of the seeds that fall to the forest floor are overlooked and allowed to sprout into seedlings.

Moreover, the timing of the masting increases the chances that newly sprouted seedlings will receive at least a modicum of rain. That's because the process that starts with flowering and ends with the fall of seed to the forest floor typically begins in Septem-ber, at the end of the dry season, and concludes in February, in the middle of the wet. Even in El Niño years, the wet season usually brings at least a few sprinkles of rain.

There was no a priori reason to suspect that the masting trig-gered by the 1997–98 El Niño would be any different. But it was. Around September, Curran and her colleagues found, the diptero-carp trees of Gunung Palung burst into bloom as expected and valiantly attempted to produce the usual superabundance of fruit.

Then something went very badly awry. Only a third of the species that produced viable seed in 1991 did so in 1998. Even more disturbing, when the researchers spread out across the park four months later, they spotted not a single new dipterocarp seedling. In 1991, the year of the last big seed crop, they had counted 155,824.

What could have caused such a spectacular failure? For one thing, the researchers observed, predation pressure on fruit and seed appeared to have increased inside the park, probably because mammals and birds were no longer able to find much to eat in surrounding tracts of logged-over land. Indeed, many of these tracts had been so extensively logged that they no longer contained any mature dipterocarp trees at all. The result: Orangutans and other animals that once would have foraged outside the park were now being forced to concentrate their food-gathering efforts within it.

There were other forces at work as well, some of them directly related to the fires that blazed in the timber and logging concessions. Among other possibilities, Curran and her colleagues suggested, smoke from the fires may have driven off or killed the insect pollinators that dipterocarp trees require in order to set their fruit. Many of these pollinators were exquisitely specific to certain species of trees, so that the species that flowered copiously but generated few fruits may have done so due to lack of pollination.

The fires very likely also contributed to the extreme dryness that prevented seeds from sprouting into viable seedlings. The fumes that rose from the burning forests of Kalimantan were laden with microscopic particles that, like little magnets, drew moisture to them. The problem was, these particles were extremely small, and much lighter than the specks of sea salt and dust that serve as the nuclei of rain droplets, and so they remained suspended in the atmosphere, creating a diffuse haze instead of clouds filled with rain.

What befell the dipterocarps of Gunung Palung was a multidimensional tragedy. Imagine that a herd of zebra on the Serengeti

Plain had the misfortune to conceive a smaller than usual number of youngsters, then the further misfortune to give birth in the midst of a gathering of lions. Then imagine that a handful of little ones somehow managed to squeeze through the predatory gauntlet only to encounter a drought so severe that they perished for lack of food and water. Add to that the idea that the lions were particularly ravenous because people had taken away their other sources of food, and that the drought was particularly severe because people had diverted the rain that otherwise would have trickled into rivers and pans. For so many reasons, then, the failure of Gunung Palung's dipterocarps to mast in 1997–98 could not be shrugged off as a simple matter of weather patterns gone awry.

THE CARBON CONNECTION

Long, long ago, around 300 million years ago, to be slightly more precise, much of the rocky substrate that presently constitutes the continents of Eurasia and North America fell within the tropics, and large areas of the world, like twentieth-century Kalimantan, were covered over with swampy lowland forests. Reptiles had just made their appearance in the world; dinosaurs, birds, and mammals were still a dream, but there were plenty of creatures rustling about. Among them, dragonflies with two-foot wingspans, millipedes as long as king snakes, and carnivorous amphibians the size of alligators.

The forests in which these animals roamed would strike us as both familiar and strange. They had dense understories of tree ferns that grew up to 30 feet tall and giant horsetails that clumped together in bamboolike thickets. Towering overhead were 150-foot-tall lycopod trees—gargantuan relatives of the lowly club mosses—that boasted thick trunks tattooed with diamond-shaped branch scars and diminutive crowns composed of grasslike leaves. When young, these trees resembled asparagus stalks poking out of the forest floor; when mature, they looked more like massive, upright logs crowned by little green parasols.

Over an evolutionary run that lasted tens of millions of years, these primitive trees and the plants associated with them left an extraordinary legacy. They were drivers of the global photosynthetic machine that helped enrich the atmosphere with oxygen and at the same time draw down its stores of carbon. This process began with the appearance of unicellular algae in the world's oceans, but it accelerated some 490 million years ago, in the so-called Ordovician Period, when the first primitive plants moved onto land. It reached a kind of apex during the Carboniferous, when vast swamps took up the carbon once sequestered by trees and transformed it into peat and then into coal.

The geochemical alchemy that performed this feat involved a number of steps and, indeed, continues to this day, particularly in Kalimantan and other warm, wet tropical places. First roots, stems, branches, trunks, and leaves sank into the anaerobic muck afforded by a network of bogs and marshes. Time passed; gradually the accumulating layers of organic material were compressed into peat. More time passed. The layers of peat were covered by sand, silt, and mud that began to weigh upon them, compressing them further. Temperatures in the compressed layers started to rise. Gradually, the peat transitioned into brown coal or lignite; the lignite, into soft bituminous coal; and sometimes the process continued to produce gleaming black anthracite.

The making of coal in the Carboniferous swamps was a lengthy process. To accumulate an inch-thick layer of peat required a decade; to accumulate a foot took well over a century. It took a layer of peat between 5 and 10 feet deep to form a 1-foot-thick seam of coal. That meant that each foot of thickness in a coal seam represented a peat bog that had been in existence for perhaps a thousand years, so that a coal deposit being mined today, in West Virginia, say, easily contains the accumulated detritus of fifty thousand or more years.

And embedded in that detritus are clues to what the climate was like when the coal seam was forming. Streaks of sediment speak of floods that swept through; fingers of charcoal testify to ancient lightning-set fires.

Until the Industrial Revolution exploded out of England in the eighteenth century, the fossilized carbon produced by ancient forests was not an important fuel. But then a Scotsman named James Watt designed an extremely powerful and versatile steam engine that changed the world. It was coal, a much more energy-dense fuel than wood, that rose to power the Watt steam engine and thus ships and railroads and industrial plants. Blanketed by horrible, sooty fogs, nineteenth-century London resembled Kalimantan in 1997–98, not because its living trees were burning—most of the great forests on the British Isles had long since been felled—but because its fossil trees were going up in flames.

Later on, coal was supplemented and in many cases replaced by oil and natural gas. Those petroleum-based fuels contained fewer contaminants and so burned "cleaner" than coal, but like coal, they represented the chemically altered remains of ancient organisms, in this case microscopic plants and animals that flourished in shallow tropical seas around the time dinosaurs walked the earth. In the 1860s, kerosene from petroleum replaced whale oil as the preferred fuel for lighting lamps. Shortly after Herman Melville finished writing *Moby Dick,* the New England whaling industry went into decline, precipitated by a number of factors, including the American Civil War and the fall in demand occasioned by the rise of the petroleum industry. Oil lamps themselves, of course, soon gave way to lightbulbs that ran on electricity, which itself was generated by fossil fuels.

In the twentieth century the pace of industrialization picked up. Around the world, in South America, in Africa, and above all, in Asia, countries rushed to catch up with Europe and North America by building factories that churned out steel and concrete, automobiles and radios, refrigerators and television sets, plastics and fertilizer, and that most useful of all commodities, electricity. As the demand for energy rose, carbon atoms that had been locked away for a hundred million or more years streamed into the atmosphere and combined with oxygen to make carbon dioxide. Imperceptibly, at first, the chemical composition of the earth's atmosphere started to change.

The biggest change occurred after the midpoint of the twentieth century. In 1958, the burning of fossil fuels added about 2.5 billion tons of carbon to the atmosphere; in 1998, it added 6.3 billion tons. Not coincidentally, over that same forty-year period, levels of carbon dioxide in the atmosphere increased from some 315 parts per million to 366 parts per million. If the trend continues, it is estimated that by the year 2050 carbon-dioxide levels might approach 600 parts per million, double preindustrial times, with potential consequences that seem well worth worrying about. With the exception of water vapor, carbon dioxide ranks as the most important of the earth's greenhouse gases.

As their name suggests, these gases resemble panes of glass that allow sunlight to stream in unimpeded while trapping a significant portion of the heat solar energy produces. It is the balance between the amount of heat retained inside the greenhouse versus the amount allowed to escape into space that keeps the earth's mean temperatures hovering at a pleasant 58 degrees Fahrenheit. If the glass were a lot thinner, the earth might come to resemble Mars, which has almost no atmosphere and shivers at minus 60 degrees Fahrenheit. If it were a lot thicker, the earth might take after Venus, a planet so encumbered by carbon dioxide that its surface temperature measures a hellish 800 degrees Fahrenheit.

Granted, Venus is closer to the sun and Mars is farther away, so these planets would need a greenhouse gas layer that, in one case, is much thinner and, in the other, much thicker to help them maintain more equable temperatures. Given its own 93-million-mile distance from the sun, the earth's mantle of greenhouse gases would appear to be neither too thin nor too thick but just right.

The role greenhouse gases play in the earth's climate system was spelled out in 1822 by Jean Baptiste Fourier, a French baron who also happened to be an extraordinarily gifted physicist and mathematician. Thirty years later, the idea that these gases were climatologically important had become very widely accepted in scientific circles. In 1858, for example, naturalist Alfred Russel

Wallace, in the course of discussing the interplay between climate and biological evolution on islands like Borneo, delivered an extremely accurate description of how the greenhouse gases work.

"The great aerial ocean which surrounds us," Wallace wrote, "has the wonderful property of allowing the heat-rays from the sun to pass through it without its being warmed by them; but when the earth is heated the air gets warmed by contact with it, and also to a considerable extent by the heat radiated from the warm earth because, although pure dry air allows such dark heat-rays to pass freely, yet the aqueous vapor and carbonic acid [carbon dioxide] in the air intercept and absorb them."

To be sure, atmospheric concentrations of carbon dioxide have fluctuated widely over the 4.5-billion-year span of the earth's history. In the very early days, for example, scientists believe, the earth's atmosphere was dense with carbon dioxide. The gas, it is thought, came from rock volatilized during fiery collisions between the young planet and the debris left over from the formation of the solar system. As a result, the earth's surface was kept quite warm even though the fledgling sun was some 30 percent less luminous than today.

Over hundreds of millions of years, the load of carbon dioxide in the atmosphere was gradually drawn down by a combination of geochemical, geophysical, and biological processes that occurred on both ocean and land. First, rain weathered the exposed rocks on the earth's surface, in the process creating a variety of carbon-based minerals. These minerals then trickled into streams and rivers and ultimately into the ocean, where they were taken up by various species of marine phytoplankton and used to make carbonaceous shells. Eventually a lot of the carbon in those shells wound up on the seafloor, where it was compressed into limestone.

Over the past billion years there have been numerous periods in which the carbon-dioxide content of the atmosphere was different from today. For example, it was quite a bit higher between 140 million and 65 million years ago, during the steamy age of dinosaurs, and quite a bit lower during the last glacial maximum. Just why this was so, however, is not completely understood.

The carbon cycle, as scientists have come to appreciate, is exceedingly complicated. Among other things, the same systems that take carbon out of the atmosphere (carbon sinks, they are called) also act as carbon sources that spew it out. The carbon that falls to the seafloor, for example, will eventually be drawn beneath the earth's crust, reprocessed into magma, and shot out by volcanoes. The waters of the ocean, for their part, alternately absorb carbon dioxide from the atmosphere and release it, depending on what the temperature is—cold water can hold more carbon than warm water—and where concentrations are greatest. Forests are an important part of the carbon cycle as well. Trees take up carbon, then die and decay and give it back, or they burn and give it back that way. Then new trees grow and take back carbon again.

How much carbon dioxide winds up in the atmosphere, then, depends on many factors. (In anaerobic environments like swamps, for example, there is not enough oxygen to produce carbon dioxide, with the result that carbon combines with hydrogen to produce another greenhouse gas, methane.)

For virtually all of earth's history, the amount of carbon dioxide in the atmosphere has been regulated by sources and sinks that were entirely natural, though in the last ten thousand years a swelling population of humans certainly contributed through activities like slash-and-burn agriculture and the burning of wood for fuel. But by the end of the nineteenth century, the large-scale switch from wood to coal had transformed human beings into a carbon source important enough to draw the attention of a Swedish chemist named Svante Arrhenius.

In the late 1890s, Arrhenius undertook a series of laborious calculations. If people continued to burn fossil fuels, he determined, they were very likely to double atmospheric concentrations of carbon dioxide, which were then approaching some 300 parts per million per unit volume of air. Such a rise, he determined, would be enough to raise the temperature of the earth by as much as 11 degrees Fahrenheit, or about as much as the extreme end of the estimates scientists would come up with a little more than a century later. Unlike his successors, Arrhenius was pleased rather than troubled by the prospect. By enhancing the greenhouse effect, he declared in a lecture delivered in 1896, the burning of fossil fuels would "allow our descendants, even if they only be those of a distant future, to live under a warmer sky and in a less harsh environment than we were granted."

FOR MANY DECADES, the idea articulated by Arrhenius—that global warming due to the buildup of greenhouse gases would be a slow-paced, pleasant affair—exerted great appeal, especially to those living in the high latitudes of the Northern Hemisphere, which was just then exiting the unpleasant cold snap known as the Little Ice Age. During this period, which lasted from about 1450 to 1850, the canals of Holland regularly froze into ribbons of ice, drifting snow marooned cattle and sheep in Scotland, and glaciers audibly groaned as they ground their way forward, terrifying farmers in the mountain valleys of Switzerland and Austria and occasionally damming up streams to create lakes. One glacier in Norway, for example, advanced 600 feet in the space of ten years.

It was this alarming advance of snow and ice, just prior to its retreat, that in the nineteenth century stimulated an impassioned scientific argument over the problem posed by the so-called errant boulders, big hunks of granite and gneiss and other hard, crystalline rock that had been mysteriously transported tens, sometimes hundreds, of miles from the geological formations to which they belonged. Quite clearly, these multiton hunks could not have traveled so far under their own power, and so scientists initially proposed that a series of catastrophic floods, including the Great Flood of Noah's day, had served as the agents of transport.

In 1833, the eminent geologist Charles Lyell began to question whether floods alone could account for such sensational displacements. He embellished the story with floes of ice. Perhaps, Lyell suggested, the boulders, rather like polar bears, had hitched rides on icebergs and ice floes that slowly drifted down from the North Pole. According to this scenario, the errant boulders found in the Alps and other mountainous regions had been stranded on high ground as the Great Flood receded.

Then, in 1840, a Swiss geologist by the name of Louis Agassiz published a tome called *Etudes Sur Les Glaciers,* which described his fieldwork on the Aar glacier high in the Jura mountains. There, Agassiz reported, he had found a field of rock so smooth that it appeared to have been sanded by a gargantuan floor polisher. Also apparent were scratches and grooves so deep that it almost seemed as if a saw had purposefully marred the smooth surface. The polisher, Agassiz proposed, was a glacier that once covered much of the region, and the scratches and grooves were the tracks of the boulders that the ice had slowly shoved along.

During a past period of intense cold, Agassiz asserted, great sheets of ice had once covered much of the known world. "The ground of Europe, previously covered with tropical vegetation and inhabited by herds of great elephants, enormous hippopotami, and gigantic carnivora, became suddenly buried under a vast expanse of ice covering plains, lakes, seas, and plateaus alike. . . . Sunrays rising over the frozen shore . . . were met only by the whistling of northern winds and the rumbling of the [crevasses]." At first, the idea seemed preposterous, but then, in 1852, an expedition established that Greenland's glaciers belonged to a solid ice cap that was but the remnant of a much vaster field of ice.

Living in a part of the world that was covered over by that ice, that indeed continued to endure very dark and cold winters, Arrhenius understandably welcomed the idea of a trend toward warmth. Besides, he could afford to remain sanguine; Arrhenius thought it would take several thousand years for atmospheric carbon-dioxide concentrations to double. As we know now, Arrhenius was wrong. Among the factors missing from his calculations was the acceleration of technological invention that would greatly increase the consumption of fossil fuels. Also missing was the exponential growth of human population made possible by better nutrition, advances in public health, and the unforeseen development of antibiotics and vaccines that beat back so many microbes responsible for deadly diseases.

The exploding number of human beings did more than increase the demand for fossil fuels; it also increased the demand for pulp and paper, for plywood and planks, for land that could be turned into houses and farms. The result was a massive assault on stores of carbon that had formerly been locked up by trees. In the twentieth century, in particular, the carbon sinks maintained by forests—living forests as well as fossil forests—were converted into carbon sources at an unprecedented pace. In 1997–98, for example, the fires in Kalimantan alone were thought to have injected some 200 million tons of carbon into the atmosphere, and the contribution from all the fires worldwide may easily have been five times that much. On a worldwide basis, it was estimated, the accelerating cycle of deforestation and fire might very well account for 25 percent of all the carbon dioxide humans were pumping into the atmosphere.

Of course, it wasn't a one-way street. Trees that did not burn but became part of a house continued to keep carbon locked up; young trees and other plants that colonized fire-scarred land took up carbon and placed it under house arrest again. Still, there was no question that the equilibrium that formerly existed between the processes that sequestered carbon and those that released it had been profoundly disturbed. Indeed, by the end of the twentieth century, carbon dioxide was accumulating in the atmosphere at such a rate that the doubling Arrhenius anticipated around the year 5000 seemed likely to occur by the year 2050 instead.

What did that mean? As scientists groped for answers, many found that they did not share Arrhenius's nineteenth-century optimism that a warmer world would be a more pleasant place. In the absence of compensatory mechanisms, the presence of additional heat seemed bound to increase the rate at which water evaporates from the surface of the earth and at the same time expand the capacity of the atmosphere to hold moisture. As a consequence, it was suggested, dry spells would be far more likely to turn into full-blown droughts, and storms would be far more likely to deliver damaging amounts of rain. Viewed through this lens, the sharp swings of the El Niño cycle came to seem like dress rehearsals for a long-running play, so that the forest fires, the disease outbreaks, the landslides that occurred in 1997 and 1998, all took on additional import.

Some climatologists, in fact, were already starting to suggest that the very modest rise in global mean temperatures that had taken place so far—1 degree Fahrenheit over the last century—had begun to spur El Niño on, causing the warm phase of the cycle to increase in both frequency and intensity. If they were right, then human beings had inadvertently harpooned the climatological equivalent of an exceedingly dangerous leviathan, one that was more than capable of rounding on its attackers and ensnaring them, and the natural world around them, in a treachery of whale lines. If they were right . . . but were they?

CHAPTER **NINE**

*Book I (Folio), Chapter I (Sperm Whale). —This whale . . .
is without doubt the largest inhabitant of the globe; the most
formidable of all whales to encounter; the most majestic in
aspect; and, lastly, by far, the most valuable in commerce.*

*Book I (Folio), Chapter II (Right Whale). —In one respect,
this is the most venerable of the Leviathans, being the first
one regularly hunted by man.*
HERMAN MELVILLE, Moby Dick

THE FARAWAY ISLAND

Turquoise, aquamarine, lapis lazuli. The ocean was the color of gemstones. The water reflected the sky, and the sky reflected the water as clouds picked up the greenish glow of the lagoon that lay just beyond the palm-fringed shoreline. Rick Fairbanks, however, was not thinking about the pas de deux that sea and sky were performing. He was immersed in the jewel-toned waters of the tropical Pacific, spiraling through them, towed by a dive boat like a lure on a fishing line.

At that particular moment, it would be fair to say, Rick had just two things on his mind. First, he wanted to find a coral colony that had started growing a hundred or more years ago. At the same time, he wanted to avoid the territorial sharks that watched over the reef as fiercely as the serpent of a thousand coils ever guarded the Golden Fleece. From time to time, he tried to turn his head, the better to spot any snaggletoothed phantoms that followed, but the force of water surging past almost tore his snorkel and mask away, forcing him to concentrate on the reef below.

Porites lutea. Montipora flagellata. Acropora hyacinthus. Mentally he checked off each coral species he spotted, at the same time marveling at how these exquisite, rocklike creatures colonized virtually every square inch of available surface save for the deep, dark canyons through which swift currents silently swept. Yet all was not well with the reef below. Everywhere he looked, Rick could see coral heads that were turning white, a sign that rising ocean temperatures had triggered an outbreak of the wasting disease called bleaching. Among coral heads that were still healthy, he spotted none that were large enough to qualify as centenarians. The sharks, too, he found underwhelming, just a couple of baby black-tips, not more than a foot or two long.

Hauling his big frame back into the boat, the forty-eight-year-old oceanographer allowed a flicker of disappointment to cross his face. Clearly he would have been happier with bigger corals, even if that had meant encountering bigger sharks.

For just a moment, Rick just stood there at the back of the dive boat, drawing deep breaths and dripping water. He was wearing faded pink swimming trunks and an old white T-shirt. His steel gray hair clung to his head like a helmet. His normally bushy eyebrows looked flatter than flower petals pressed between the pages of a book, and his arms and face were mottled with freckles, the legacy of years spent in open water. As he tossed his mask and snorkel to one side and pulled off his electric blue fins, he remarked to no one in particular, "I'm afraid that we're just too close to the signal; El Niño is killing off the messenger we seek."

Rick was referring to the fact that reef-building corals are the biological equivalents of thermometers and rain gauges, which nature, for its own inscrutable reasons, has seen fit to string out across tropical waters. From living corals he and others had extracted climate records that in some cases were several centuries long; from fossil corals, they had reconstructed climate records that sometimes leapfrogged across millennia. In this case, Rick was hoping to come back with specimens from both the quick and the dead. The problem was, the powerful El Niño that had surfaced some six months earlier kept interfering with his plans.

It was November 1997, almost five weeks to the day that Rick had left his laboratory at Columbia University's Lamont-Doherty Earth Observatory and headed out to Honolulu. With him he brought more than 30 tons of gear. The dive boat. A sophisticated sonar system for surveying the island's submerged terrain. Dozens of glass vials for taking water samples. Chemicals for performing experiments. Oceanography texts packed with equations and charts. Several field guides to Pacific corals. And last but not least, an industrial-strength drill from the Texas oil patch for coring into ancient reefs buried beneath the seafloor.

After stowing all that aboard the *Moana Wave*, a research vessel operated by the University of Hawaii, Rick and about a dozen colleagues set off for the central Pacific atoll of Kiritimati, which lay some 1,200 nautical miles to the south. As it happened, the *Moana Wave* was the same ship that twenty-two years earlier had plied up and down the coast of Peru in an attempt to intercept the El Niño that oceanographer Klaus Wyrtki so fervently believed was forming. While the *Moana Wave* had not succeeded on that go-round, it was more than making up for it now.

Rick had chosen Kiritimati as his focus because its location—a tad north of the equator, a jot east of the International Date Line—placed it directly in El Niño's path. What he hadn't counted on was the challenge of prospecting for corals in the middle of one of the greatest El Niños to surge across the Pacific in one hundred, three hundred, perhaps even a thousand years.

The first signs of trouble manifested themselves on the way down from Hawaii, when the *Moana Wave* set anchor off the island of Palmyra just as a fierce squall was blowing up. Seabirds by the hundreds swarmed the ship, taking refuge in every conceivable nook and cranny. Then waves started crashing over the fantail. One was a 10-footer that rolled all the way up the deck, sweeping several of the scientists off their feet and smashing Rick against a railing. For days afterward he swilled ibuprofen to kill the pain in his spine. Such storms were not usual in this part of the ocean, he observed, which lies in the so-called dry zone of the equatorial Pacific.

And so it went for several weeks on end.

Even when there were no local squalls roiling the waters, the ocean often enough turned surly. Because of its equatorial location, Kiritimati sat right in the firing line of storms that roamed both hemispheres, so that depending on where the *Moana Wave* was anchored, it could come under attack from waves produced by a storm in either the North or the South Pacific. In late October one such storm swell had caused the ship to lurch so violently that the 20-foot-tall derrick on the deckside drill cracked and started to topple over. The force of the fall ripped the "R" off the name emblazoned across one of the topmost struts. Instead of ACKER, the drill's manufacturer, the steel letters spelled out ACKE.

The Acker was a quasi-portable drill used by petroleum geologists. It was much lighter than most oceangoing drills but rugged enough to do the job, or so Rick was hoping. Before going to sea, he and his colleagues had tested the drill in the Lamont-Doherty parking lot, where it performed splendidly, cutting through thick layers of asphalt and bringing up neat cylinders of dirt and rock. The problem was, unlike the parking lot, the *Moana Wave* kept lurching up and down. Now Rick was wondering how much longer ACKE would last.

ON A TOPOGRAPHIC MAP, Kiritimati resembles a stocking-shaped stepping-stone that protrudes just barely above the surface of the sea. Its highest point, a knobby outcropping known as Joe's Hill, stands not much more than 40 feet above sea level, and the rest of the place ranges from a minimum elevation of 5 feet to a maximum of 12. In all, the island boasts some 150 square miles of solid ground, a seemingly unimportant fact that nonetheless enables it to claim the title of world's largest inhabited coral atoll. Among the local landmarks that unfailingly impress is the watery graveyard known as the Bay of Wrecks, where a whaling ship named the *Briton* went down in 1836, followed by the *Mozart* and the *Maria Helena* some years later.

Like other motes of land that dot the great oceans, Kiritimati appears to have been discovered not once but several times, first by Polynesian voyageurs (who called it the Faraway Island), then by the Spanish, and finally by Capt. James Cook, who, on December 24, 1777, sailed in from the northeast, toward the mouth of the lagoon, bestowing on it the name that stuck, Christmas Island. When Christmas Island became part of the island nation of Kiribati in 1979 (the "ti" is sounded as an "s"), the spelling of its name changed to reflect the conventions of the Micronesian language its citizens adopted.

The change in spelling was not quite so arresting as the change in the calendar, which advanced by exactly one day. For while a quirk of geography placed Kiritimati in the same time zone as Hawaii, a quirk of politics required its residents to take an imaginary leap across the International Date Line, where the bulk of the Republic of Kiribati resides. In all, thirty-four atolls and islands make up this mid-ocean nation, including Kanton, formerly Canton, whose oddly punctuated swings from drought to rain had given Jacob Bjerknes such a priceless clue to El Niño some forty years earlier.

These days Kiritimati has a population of around four thousand, mostly immigrants from the nearby Gilbert Islands. It also boasts a wildlife sanctuary that shelters some eighteen species of seabirds and a languid lagoon, which, according to those who know about such things, qualifies as one of the finest bone fisheries in the world. Open to the sea on one side, this intensely blue, intensely saline lake is continuously refreshed by tidal surges; in addition to bonefish, other sea creatures, including a goodly number of stingrays, pass in and out. Sharks make the passage, too, though not in the threatening numbers that Captain Cook and his men reported.

In 1777, for example, James Trevenen, a midshipman on Captain Cook's ship, the *Resolution,* rowed into the lagoon to hunt for turtles. He wrote:

> On every side of us swam Sharks innumerable and so voracious that they bit our oars and rudder and I actually struck my hanger two inches into the back of one whilst he had the rudder in his teeth. The boats fishing . . . in shallow water, carried long pikes to keep the Sharks from the bait. On board the ships the Sailors caught great numbers, and . . . contrived a hundred ways to torment them, sometimes two were firmly lashed together by Tails and turned adrift—others had large pieces of board tied under their tails.

Seabirds, attracted by an abundance of crustaceans, squid, and small fish, were also plentiful in Cook's day, among them, boobies, sandpipers, plovers, curlews, and terns. But other than "some cleared places such as where they thought huts had been," the crew of the *Resolution* detected no signs of prior human habitation. This did not surprise Cook, who commented on the contrast between the richness of the ocean and the poverty of the land. The island, as far as he could tell, had "no visible means of allaying thirst, nor any vegetable that could supply the place of bread."

Once, in fact, two of Cook's men lost their way amid the island's low bushes and, after hours, found themselves so desperate for water that they resorted to drinking the blood of turtles and birds. Eventually, both were found, one of them "almost dead," according to one of Cook's officers. "Having met with a pool of clear water he went into it with his clothes on in hope of refreshing himself, but unluckily this water to which he had been invited by its transparency proved no other than a pool of strong brine, by which his limbs and clothes became so stiff that he was hardly able to crawl out."

The reason Cook's men became so hopelessly lost became clear when Rick led an onshore expedition that involved a quarter-mile swim through pounding surf to a nearby beach. The objective of the outing was to gather up coral boulders that had been cast up by storms; Rick thought that one or more might contain a record of El Niño that went back not a century or two but several thousand years. The job of gathering up the specimens didn't take long, as the boulders were strewn everywhere. They littered the coastline. They sat in puddles. They even hid out under bushes, along with dozens of hermit crabs that huddled together in shelly camaraderie.

As Rick and a small party ventured farther inland, however, they—like the members of Cook's crew—soon lost sight of the ocean. Aside from a narrow tarmac road and a smattering of scrub, the terrain looked almost featureless. A small lagoon appeared. As Rick knew, it was only one of many that turned the interior of Kiritimati into a treacherous maze. "Normally, these ponds are so filled with shrimp that they're rusty red in color," Rick remembered from a past visit, "and the birds just sit there and stuff themselves." This time, though, there didn't seem to be any brine shrimp; there didn't seem to be any birds.

Kiritimati, as Rick found it in October and November 1997, was strikingly different from the place Captain Cook had described. By that point, the temperature of the ocean in the vicinity of the island had shot up as much as 9 degrees Fahrenheit above normal, and the great rainmaking machine that usually sat over the Indonesian archipelago was now positioned directly overhead. All around, the buoyant mist exhaled by the ocean condensed into puffy white plumes that shot skyward like geysers. Downpours drenched the island almost daily, flooding the streets that ran through sleepy fishing villages named Banana and Poland and changing the coloration of the palms along the shoreline from burnt orange to exuberant green.

For a time, it was hard, even for those who lived there, to remember that Kiritimati was normally an extremely arid place whose only surface water consisted of salty, shrimp-filled ponds.

The effects of El Niño on Kiritimati are very similar to its effects on coastal Peru. As the waters of the central Pacific warm, squid and small fish become scarce, while on land, flooding rains wash out the eggs laid by seabirds and the fledglings in their burrows and nests. During the 1982–83 El Niño, for example, millions of adult seabirds staged a mass exodus from Kiritimati. Only a fraction returned.

Two ornithologists from the Los Angeles County Museum of Natural History, Ralph and Elizabeth Anne Schreiber, chronicled their travails in the journal *Science*. The population of great frigate birds, they reported, experienced total reproductive failure in November 1982. "Approximately 90 nestlings should have been present in our study colonies," they observed, "but we found only six still alive, none of which was able to fly and all were starving." Many of the adult birds eventually died as well. A year later the Schreibers found only six hundred breeding great frigate pairs, down from their previous count of six thousand.

The bleaching of the corals off Kiritimati was another of El Niño's characteristic calling cards. Corals in this sector of the ocean had adapted to water temperatures that ranged between 70 and 79 degrees Fahrenheit, and when temperatures shot into the eighties, as they did in 1997, they started to turn deathly white.

Individually the polyps that make up a coral head are carnivorous animals akin to sea anemones and jellyfish; they feed at night on microscopic animals, which they stun with poison-tipped tentacles. But the polyps also derive a good bit of nourishment from sugars produced by little photosynthetic algae that dwell in their soft tissues. These tiny symbionts are known as zooxanthallae, and it is their presence that lends coral heads their muted earthtone coloration. When the "zoos" leave, the polyps become transparent, exposing their white exoskeletons.

Bleaching appears to be a generalized response to stress, including thermal stress, though just why it occurs remains something of a mystery. Some scientists believe it's the coral polyps that expel the zoos; still others think that the zoos reject the polyps. Both processes could well be involved just as transplanted tissues and organs sometimes reject patients as well as the other way around. Like the flushing that accompanies a fever, bleaching is symptomatic of an underlying disorder, and depending on how long it goes on, it may or may not prove fatal. During the wave of bleaching caused by the 1982–83 El Niño, reefs across the eastern basin of the tropical Pacific lost many of their largest, most spectacular corals.

As TIME WENT ON, the odds of finding a living centenarian coral in Kiritimati's disturbed waters began to seem increasingly slim. Yet even so the search continued, powered by four divers from the University of Hawaii whose enthusiasm for getting wet seemed boundless. On one occasion, the fearless foursome found a coral head of such promising size, such generous proportions, that they decided to float it up to the surface, producing a graceful underwater ballet that lasted about ten minutes. As they worked some 60 feet beneath the surface, they exhaled streams of fizzy bubbles, so that it looked as if they were swimming through soda water.

First they wrapped the coral in a cargo net; then they attached balloonlike lift bags to the net and inflated them with air, using a spare regulator. As the coral gained buoyancy and began to rise on its own, the divers guided it to the surface and used the dive boat to tow it back to the ship. The *Moana Wave*'s big crane then winched the coral from the water and set it down gently onto the upper deck. Seawater gushed from a myriad of crevices and crannies, and the polyps that covered the coral's exterior exuded a fishy smell. Startled by the sudden influx of air and light, a brittle star crawled out of its hiding place, followed by a tiny crab, which clambered over the barnacles growing on the coral's pedestal.

Methodically, Rick scanned the coral's surface, first determining its species, then inspecting it for structural defects, like holes made by boring sponges. At that point, he tried to gauge its approximate age. Alas, he concluded, this particular coral was probably an octogenarian at best. Water, as he well knew, acts as a magnifying lens, and so to the divers the coral head had looked much larger, and therefore older, than it actually was. "It's close," he said with a wry smile. "But next time, when I say three meters, I really mean three meters."

What Rick was dreaming of finding at Kiritimati was a proximate match to the prize an earlier expedition to the Seychelles had turned up the year before. Just offshore Mahe Island, in the southwestern sector of the Indian Ocean, a team of scientist-divers associated with his lab at Lamont-Doherty and the Scripps Institution of Oceanography had happened across a magnificent coral head lurking beneath some 20 feet of iridescent water. It belonged to the species *Porites lutea,* and it had attained the impressive height of 3 meters, or around 10 feet. The scientists later confirmed what the size of the coral head suggested—that it was around two centuries old.

Like other reef-building corals, the Seychelles *Porites* had begun life as a single polyp that was smaller, even, than a baby's fingernail. From the surrounding seawater, the polyp had extracted the building blocks for making aragonite, a chalky mineral from which it fashioned a protective cup. Soon the founding member of the colony divided, becoming two polyps, and those polyps divided in turn. Each new generation of polyps made a new generation of aragonite cups, which joined to form a sturdy scaffoldlike structure. In this fashion, the *Porites* slowly grew both vertically and horizontally, like a mighty fir or spreading oak, with a veneer of living tissue a fraction of an inch thick covering an ever-enlarging core of stony heartwood.

As it grew, the Seychelles *Porites* meticulously laid down tree-like growth rings, with the consequence that the layers of aragonite it produced were arrayed in clear temporal order. These layers were the chalkboards on which successive generations of polyps inscribed a record of the conditions under which they had grown. When the temperature of the water rose, the polyps responded by slowing the rate at which they incorporated certain minerals and other geochemical compounds into their hard, mineralized shells. When sea-surface temperatures fell, they performed the same trick in reverse. In this fashion, the coral head had gradually assembled a climatological archive that was nearly equal in quality to the instrumental record, only longer.

The instrumental record of tropical climate goes back to only about 1880 where the atmosphere is concerned and only the 1950s in the case of the oceans; furthermore, there are gaps in the record in important areas of the world that even today are not well covered by instruments. Corals had the potential to fill in the gaps in the modern record and extend it significantly in time. Fossil corals could open windows that sometimes went back more than one hundred thousand years. Living corals could probe the more recent past, sometimes doubling or even trebling the length of the instrumental record.

Massive corals more than three hundred years old had been found growing off the Galápagos Islands; corals more than five hundred years old were known to inhabit the Great Barrier Reef. Granted, Rick did not expect the highly disturbed waters off Kiritimati to harbor a coral quite so venerable. But he did think that it might be possible to find a colony that had started growing more than one hundred years ago, so that like the Seychelles coral it would contain a record of sea-surface temperatures that went back to 1877, the year in which a powerful El Niño coincided with a catastrophic failure of the Indian monsoon.

Startling though it seemed, Kiritimati's transformation in 1997 was an expected consequence of El Niño. The climatological condition of the Indian subcontinent was another matter, and it was very much in the background of Rick's mind as he prospected for that centenarian coral in Kiritimati's waters. In the 1980s the weakening of the Indian monsoon in response to El Niño had seemed like a textbook example of the far-reaching influence of the tropical Pacific. Indeed, the comparatively moderate El Niño of 1986–87 coincided with the worst drought the Indian subcontinent had experienced in something like a hundred years. But in the summer of 1997, in the face of one of the most powerful El Niños ever, India received an almost perfectly average amount of rain.

"Sometimes El Niño and the monsoon play together," Rick reflected. "Sometimes they play separately. What I want to know is, where is the conductor?"

The Seychelles *Porites* strongly suggested that scientists were still missing an important piece of the puzzle. For two hundred years, the *Porites* had grown in a sector of the Indian Ocean that strongly experiences the wind shifts associated with the Indian monsoon, and over that time period, it had recorded rhythmic swings in sea-surface temperatures that could be sorted into four highly distinctive beats. There was a high-frequency beat that corresponded to seasonal fluctuations in ocean temperatures; there was a lower-frequency beat that reflected year-to-year fluctuations due to the monsoon. And then, superimposed on top of those, were two other beats, one that had a frequency of three to six years and another that averaged out at around twelve.

In the first instance, the sea-surface temperature fluctuations mirrored the back-and-forth swings of the El Niño cycle, leaving little question as to the identity of the source. But whence did the twelve-year signal come? The monsoon? El Niño? Some other climate cycle, as yet dimly glimpsed and imperfectly imagined, that affected them both?

CYCLES WITHIN CYCLES _____

In the 1920s, physicists had an appealingly simple picture of subatomic structure. Positively charged protons and neutrally charged neutrons occupied the nucleus of the atom, which was enveloped by a fuzzy cloud of negatively charged electrons. Then, in 1936, a fourth type of particle appeared in a detector that was gathering up cosmic rays like a fisherman's net. At first, physicists thought the particle was the pion, whose existence theorists had postulated on the basis of the need for some sort of nuclear glue to hold protons and neutrons together. It wasn't until some time later, shortly before the pion itself showed up, that the strangeness of the particle discovered in 1936 became apparent.

That particle was the muon, a heavier cousin of the electron, and it provided the first strong hint that the subatomic world was far more complex than physicists had theretofore imagined. "Who ordered that?" Columbia University physicist Isadore Rabi is said to have exclaimed in surprise, and much the same question could have been asked about El Niño. No less than the muon, El Niño greatly complicated the prevailing view of why climate changed and how. It was what alerted scientists to the importance of the internal resonances the climate system was capable of conjuring into being. It was what made them think that there might be many more such resonances, with time scales that ranged from decades to centuries.

The old view of climate embraced the twenty-four-hour diurnal cycle, the twelve-month seasonal cycle, and the long, ponderous ice-age cycles, all of which were driven by fluctuations in the distribution of sunlight across the surface of the earth. It also embraced the more tentative idea that there might be cycles driven by variations in solar luminosity, such as the eleven-year sunspot cycle. And then there was the Asian monsoon, which fit into the seasonal cycle as neatly as the electron fit into the atom and yet, in retrospect, afforded the first indication that something more might lie beyond. That's because the Asian monsoon, no less than El Niño, is an oscillation of the climate system.

Indeed, the Asian monsoon is often compared to a giant oscillating sea breeze that sets up in response to temperature contrasts between the surface of the land and the surface of the ocean. Land and water have different thermal properties; in summer the former heats up faster than the latter, and in winter it cools down more rapidly. In both cases, a temperature gradient is set up that causes air to move, in one case from the ocean onto the land and in the other case from the land onto the ocean.

It is for this reason that coastal zones and islands like Kiritimati are regularly swept by refreshing afternoon sea breezes. Starting at dawn, the sun beats down on exposed sand and rock, causing them to warm. The heat rising from the land in turn causes the overlying air to warm and therefore to rise. Into the void created by this rising air rushes the slightly cooler air that hangs over the ocean. Then, at night, as the land cools down, the flow pattern reverses.

The Asian monsoon system has two main arms: the Indian or South Asian monsoon and the East Asian monsoon, which is known as Mei-yu in China, Baiu in Japan, and Chang Mau in Korea. There are numerous regional differences. In China, for example, the summer monsoon starts in the south in late spring, progressing to northeast China by late summer. In India, the summer monsoon typically begins around the first of June and very quickly spreads; two weeks later, it encompasses virtually the whole of the country.

Both arms of the Asian monsoon work on the same basic principle. Consider what happens on the Indian subcontinent. In the winter, a great blast of northeasterly winds comes swooping down from Siberia, carrying cold, dry air across the Himalayas toward the warm waters of the Indian Ocean. In the summer, as land temperatures rise, this flow reverses so that the winds approach from the Arabian Sea and the Bay of Bengal, laden with the marine moisture that will produce the miracle of rain.

What drives the wind reversal is the tremendous amount of radiation that the summer sun dumps onto the Qinghai-Tibetan plateau. This big plateau, the world's highest, is rimmed by the Himalayas, the Karakorum, and the Tien Shan mountains; the Ganges and the Yangtze Rivers start their journeys there. In all, the Qinghai-Tibetan plateau covers an area nearly half that of the United States and boasts a mean elevation of more than 3 miles, making it thrice as high as Denver and even more exposed to intense solar radiation.

In the wintertime, needless to say, the Qinghai-Tibetan plateau is an icebox. Starting in spring, however, it very swiftly heats up, and so does the air above it. As the air warms, it rises, creating a zone of low pressure into which marine air then rushes. In a low-lying state like Florida, the inflow of air from the ocean often produces thunderstorms on a summer afternoon; in India, where warm, moist air is forced up and over the Himalayas, the effect is magnified manyfold. The monsoon endows the city of Cherra-punji in mountainous Assam with an annual allotment of 426 inches of rain, a distinction that ranks it, alongside some obscure location in Hawaii, as the rainiest spot on earth.

Given the power of the monsoon system, perhaps it should not be surprising that it does not always respond to El Niño in a predictable fashion. On the one hand, the very worst monsoon failures on the Indian subcontinent have almost invariably coincided with El Niños; on the other hand, the converse does not always hold. That is, while the monsoon often perceptibly weakens in El Niño years, not every El Niño brings India a shortfall of rain. From the late 1800s until around 1920, for example, the monsoon appears to have been very closely synched with the El Niño cycle, faltering, often dramatically so, when the tropical Pacific warmed, rebounding when it cooled. Then, between about 1920 and 1960, the monsoon seemed to assert its independence.

This same mid-century time period, however, was also marked by a rather sparse sequence of El Niños, so that the question naturally arose as to whether it was some weakness in the El Niño cycle that prompted the monsoon's strength or whether it was the other way around. Just as El Niño is capable of suppressing the monsoon and La Niña is capable of energizing it, so it had long appeared, to some scientists at least, that the monsoon might be equally capable of influencing El Niño. As Sir Charles Norman, one of Sir Gilbert Walker's protégés, observed back in 1953, "The monsoon stands out as an active, not a passive, feature in world weather, more efficient as a forecasting tool than an event to be forecast."

In the 1990s the monsoon was strong despite the influence of El Niño.

BY THE END of the twentieth century, climatologists trying to understand why El Niño's impacts were so variable found themselves contemplating the possibility that still other oscillations in the climate system might play a mediating role. Virtually all of these oscillations bore acronyms that ended in a telltale "O." There was the QBO, or quasi-biennial oscillation, which involves the reversal of stratospheric winds over the tropics every twenty-six months or so; the TBO, or tropospheric biennial oscillation, a kind of lower atmosphere equivalent that affects ocean temperatures and wind speeds in both the Indian and Pacific Oceans; and the MJO, or Madden-Julian oscillation, which triggers an extremely broad band of convective activity that sweeps from west to east across the equator every thirty to sixty days.

And as if that weren't enough, there were at least two other oscillations of importance—the NAO, or North Atlantic Oscillation, and the PDO, or Pacific Decadal Oscillation, both of which, like the El Niño Southern Oscillation cycle, were originally defined purely in terms of their atmospheric manifestations.

The oscillation in the North Atlantic involves swings in the strength of the low-pressure system located around Iceland. In years when the low is strong, the so-called positive phase of the oscillation, a big blast of frigid continental air sweeps across the northern tier of North America, then out over the Atlantic Ocean, where it picks up warmth from the waters below. As a result, northern Europe experiences milder winters, while Greenland and Labrador shiver under a coverlet of cold. Then, sometime later, the negative phase of the cycle kicks in, and the pattern reverses. As Greenland and Labrador warm up, northern Europe grows chill.

The atmospheric side of the oscillation in the North Pacific is similar, though it goes by a slightly different name: the Pacific North American pattern. Just like its North Atlantic cousin, the oscillation in the Pacific has two phases, each reflecting the relative strength of the low-pressure center around the Aleutian Islands. Shifts in the strength (and also the position) of this low-pressure center affect the flow of weather across the whole of the North American continent. In one phase, the winter storms that hit the United States are more likely to follow a southerly track, so that winters in California turn wetter and those in the

Pacific Northwest turn drier. In the other phase, conditions essentially reverse.

Initially it was assumed that the oscillations in the North Atlantic and the North Pacific switched from one state to another on a purely random basis, so that they were coherent in space but not in time. In the 1990s, however, as scientists scrutinized the behavior of these switches over longer time scales, they began to convince themselves that they saw distinctive long-term patterns. The pressure seesaw in the North Atlantic, they observed, had been heavily biased toward a negative state between about 1950 and 1970 but then switched to a more strongly positive mode. The pressure seesaw in the North Pacific likewise shifted its preferred modus operandi around the same time.

The length of the time frames involved put intense focus on the role that it seemed only the ocean could play, largely because, as noted before, unlike the atmosphere, the ocean has a long-term memory. The reason has to do with an essential difference between the two media. Molecules in the atmosphere are in a gaseous state and so move around very quickly. Winds blow, then stop; air warms, then cools; storms form, then dissipate. In the space of a few days, the information the atmosphere contains about such things as temperature and pressure and moisture is all but completely erased and overwritten with new information. The molecules that make up the ocean, by contrast, are packed much more closely together and, as a consequence, their ability to move around is more tightly constrained. By volume, it is worth noting, a unit of water weighs a thousand times more than air. As a consequence, things that happen to the ocean—changes in sea-surface temperatures, for example—leave imprints that can last anywhere from decades to hundreds of years.

For this reason, then, scientific attention soon focused on the great windblown gyres in both the North Atlantic and North Pacific. Gyres, in essence, are just circular current flows that shunt warm water from the equator to the poles and cold water from the poles equatorward; as a result, higher-latitude waters do not grow too cold, nor do tropical waters become too warm. The eastern arm of the subtropical gyre in the North Pacific, for example, is the cold, south-flowing California Current; the western arm, the warm, north-flowing Kuroshio Current. In similar fashion, the subtropical gyre in the North Atlantic includes the warm, poleward-flowing waters of the Gulf Stream and the cool, south-flowing waters of the Canaries Current.

The gyres, a number of scientists suggested, could easily rearrange patterns of sea-surface temperatures simply by spinning big blobs of warmer and colder water around, and in the twentieth century, at least, this was what appeared to be happening. In the North Atlantic, oceanographers tracked two such blobs as they completed one full turn of the subpolar gyre, starting out in the Gulf Stream, then ending up, about two decades later, in the Labrador Sea. One, which was unusually warm, started on its way in the early 1950s; the other, which was anomalously cold, set off in the late 1960s.

In the North Pacific, scientists identified no less than four highly suspicious shifts in sea-surface temperature patterns in the twentieth century. A shift that occurred around 1946, for example, brought cooler-than-normal sea-surface temperatures to the west coast of North America and warmer-than-normal sea-surface temperatures to the central Pacific; a shift that occurred in 1976 imposed an opposite pattern.

There was little question that the shifts themselves were real. The problem for scientists was to elaborate a mechanism capable of generating and sustaining these shifts as part of an ongoing oscillation. Symptomatic of the lack of clarity was the fact that scientists often resorted to body language when attempting to explain their ideas to others, entertainingly windmilling their arms in imitation of the great circle routes the ocean gyres follow. As with El Niño, they argued, the currents responded to the winds, and the winds, in turn, responded to the sea-surface temperature patterns generated by the currents. But which was the chicken, which the egg?

For the interdisciplinary field of climatology, the 1990s were both an enormously exciting and an extremely confusing time.

Nothing was certain; everything was in flux, including the defining characteristics of the phenomena scientists were seeking to explain. Right after the El Niño of 1997–98 ran its course, for example, scientists from the University of Washington would advance the case that the North Atlantic Oscillation was just a branch of an even larger pan-Arctic oscillation. In its horizontal dimensions, this oscillation formed a bridge between the North Atlantic and the North Pacific. In its vertical dimensions, it stretched from the stratosphere to the surface of the ocean.

The Pacific Decadal Oscillation, for its part, was also in conceptual evolution, growing from a phenomenon confined to the North Pacific to one that embraced the South Pacific and the tropical Pacific as well. Some speculated that the warming and cooling of the tropical ocean was being driven by events in the higher latitudes, while others argued that it was more likely the other way around. Or perhaps both sides were right, so that the back-and-forth between the equatorial Pacific and its higher-latitude coconspirators constituted yet another oscillation waiting to be discovered. If that proved to be the case, then the opportunity for unexpected interactions between oscillations would exponentially expand.

Now things were starting to get really complicated. Like musical instruments, multiple climate cycles seemed to be overlapping in time and space in various ways. If several out-of-tune instruments—a violin, a cello, and a piano—try to play an A sharp simultaneously, for example, the sound will be perceived as an annoying quaver. By contrast, if all the instruments are in tune, then the A sharp will be one clear tone. And if the same in-tune instruments simultaneously play three different notes, they may produce a richly harmonious chord or a jarring dissonance.

In the case of musical instruments, this happens because sound is made of waves whose frequencies overlap to produce different effects. Climate cycles, too, have characteristic frequencies—the time intervals at which they repeat—and when two or three or more overlap, they make their own strange music. As the El Niño of 1997–98 exploded across the world stage, scientists were twiddling with their statistical tuners, straining to hear that music. They were trying to figure out what the different instruments in the climate orchestra were, what score these instruments were following, and whether they were even reading from the same page of music or not.

BACK TO THE FUTURE _____

It was about nine at night. Rick was standing on the deck of the *Moana Wave,* along with several other scientists. Save for the occasional 2-foot swell sweeping toward shore, the ocean had gentled, and as moonlight spilled across its onyx surface, Tony Castillo, a professional oil driller from Texas, stood on a platform suspended over the railing of the ship. Beneath Tony's feet, the silvery forms of impressively large fish occasionally streaked very close to the surface. Some of them were the fabled bonefish, and the water through which they swam lapped at the *Moana Wave* like wavelets approaching the edge of a lake.

The reef into which Tony was trying to cut lay some 200 feet below the sea's surface. It had, in essence, "drowned" as sea levels rose between thirteen thousand and eight thousand years ago, and the corals that built it were no longer alive, but formed thick layers of aragonitic limestone separated by lenses of sand. As the drill bit hit sand, then rock, then sand again, it often skittered, and even worse, sometimes got stuck. The process of freeing it, then working it back into position, could take a couple of hours, sometimes more. On this particular night, though, things were going well, and Tony repeatedly brought up cylindrical lengths of coral as white as bleached bone. "This is the good stuff," Rick exulted. "This is what we came here to find."

It was a labor-intensive operation. Tony continuously had to extend the drill's range by adding new sections of 20-foot-long steel pipe. To reach the seafloor required twelve sections. Then, as the drill cut deeper in, new pipe had to be added. Like pieces of a giant Lego set, the sections lay in a big pile on deck, and they had to be placed in Tony's hands. Because of his back injury, Rick was exempted from this particular labor, but two of the other scientists on board were putting in the requisite twelve-hour shift, working as roustabouts, "slinging pipe" for Tony. Their eyes glazed over with fatigue, and under a layer of grease and sweat, their skin shone pale, as if the effort had drained them of blood.

The drilling site lay near the mouth of the lagoon, near Vaskess Bay, on the western side of the atoll, and its location had been dictated by sonar surveys of Kiritimati's submarine topography. This was the one place that provided a broad enough platform for the drill's bit to gain a purchase, as below its low-lying surface, Kiritimati was anything but flat. It had been born, as coral atolls are, when, many millions of years ago, a steep-sided volcano raised itself up from the ocean floor. Over time, coral larvae, borne by ocean currents, settled on the submerged flanks of the volcano. The larvae grew into polyps, the polyps into colonies, and the colonies into a reef that, over time, built itself into an atoll whose lagoon filled in the old caldera.

During successive glaciations, as sea levels fell, the corals migrated down the atoll's precipitous slopes. Then, as the glaciers melted and sea levels rose, the corals climbed back up again. This penchant for marching up and down stemmed from the fact that reef-building corals are denizens of the translucent shallows, drawn to the sun for a very good reason. Like all green plants, the little symbiotic algae that help sustain the corals cannot live without sunlight, which does not penetrate much more than a couple of hundred feet below the surface. Moreover, the corals in the drowned reefs the drill was probing belonged to a shallow-water species; when they were alive, they grew in water no more than about 20 feet deep.

These corals, Rick hoped, would contain a record of the swings in sea-surface temperature that occurred during an interval of time that began roughly thirteen thousand years ago and ended some twenty-five hundred years later, when the Northern Hemisphere, at least, experienced an extremely sudden snap into cold conditions, followed by an astonishingly abrupt lurch into warmth. There was now accumulating evidence that this climatological upheaval had been global in scope, and so a picture of how this region of the Pacific had behaved during this time would be extremely valuable. Moreover, if Rick was really lucky, the corals he retrieved might contain a record of how El Niño had responded as well.

The most common way of thinking about El Niño is as an ocean sloshing that warms the waters of the eastern Pacific. But another way, the way Rick often thought of it, is as a wandering of the large-scale atmospheric feature known as the Indonesian Low. The Indonesian Low is one of three great convection engines situated along the equator—"the fireboxes" of the climate system, many term them. The other two are positioned over the densely forested drainage basins of the Congo and Amazon Rivers.

The great cloud plumes that rise from these hot, moist regions do much more than supply rain to the landmasses below. They also transfer the solar energy absorbed by the earth's surface into the high reaches of the atmosphere, thereby speeding the transfer of heat from the earth to space and from the equator to the poles. When molecules of liquid water evaporate to form vapor, they carry away little payloads of excess heat; when they condense into cloud droplets, these payloads explode, leaving bursts of heat behind. Like the great ocean gyres, the equatorial fireboxes perform a vital function: They keep the tropics from growing too hot and the poles from growing too cold.

What Rick found intriguing was this: Only one of the fireboxes—the Indonesian Low—repeatedly ran away from home. After all, here it was, in November 1997, creating puddles in Kiritimati when it should have been back in Borneo keeping the rain forests wet. Why would the Indonesian Low behave in so deviant a fashion?

Zones of intense convection, Rick thought, tended to be positioned over the continents rather than over the oceans because land, particularly dark, forested land, heats up so much faster than water, and it is heat, after all, that causes air to rise. Today, perhaps, the Indonesian Low did not have quite enough land to keep it firmly pinned down. But during the peak of the last glaciation, when sea levels were an astounding 400 feet lower, much more landmass would have been exposed.

Not so very long ago, it was possible to walk from Australia to New Guinea across the Torres Strait; it was also possible to walk across the Sunda Strait from Borneo to Sumatra. Perhaps El Niño emerged from semihibernation around eleven thousand years ago, Rick speculated, as these low-lying land bridges started to sink beneath the waves.

EL NIÑO, the Asian monsoon, and all the other oscillations are features of the present climate system. They did not exist 4.5 billion years ago when the earth emerged from its chrysalis of rock and dust and settled into its quotidian spin. They also did not exist, at least in their present form, 65 million years ago, when a large asteroid or comet hurtled in from outer space and ended the long reign of the dinosaurs. That's because, before the music could begin, nature had to sculpt the instruments on which to play it, in the case of El Niño, an equatorial ocean basin of the right shape and size.

The first step in the process can be traced back to the breakup of the supercontinents that once held the present-day continents in a constrictive embrace. At one point, for example, there was a single supercontinent named Pangaea; then Pangaea broke into two supercontinents, Gondwanaland and Laurasia. To Gondwanaland belonged what was to become South America, Africa, Australia, Antarctica, and India; to Laurasia, North America and much of present-day Eurasia. Between these great landmasses lay a vast sea called the Tethys, which, at one point, snaked its way almost completely around the world. The Mediterranean Sea is a fossil remnant of the Tethys.

The big breakup started roughly 160 million years ago and took a very long time as great landmasses ponderously collided and rifted apart. Eventually, continent-size chunks of the earth's crust irrevocably separated, slowly sailing away from one another over slithery expanses of semimolten rock. A big piece of Gondwanaland slowly crunched into central Asia, forcing up the Himalayas. Africa and South America separated, likewise North America and Eurasia. Australia and New Guinea trended north, progressively narrowing the connection between the tropical Pacific and Indian Oceans.

These changes profoundly affected climate. For example, the fact that Antarctica separated from Gondwanaland and ended up at the South Pole set the stage for a rhythmic succession of glaciations in the Southern Hemisphere. The poleward wending of North America and Eurasia did the same for the Northern Hemisphere. In similar fashion, it was the rise of the Himalayas that helped jump-start the Asian monsoon; the rise of the Andes that cast a rain shadow over the desert coast of South America; and the rise of the Isthmus of Panama that blocked the outflow of warm tropical water from the Atlantic Ocean, thereby energizing the Gulf Stream.

Last but not least, it was the formation of the vast ocean basin in the tropical Pacific that created the incomparable instrument that El Niño and La Niña so masterfully play. In this case, the sea's surface is like a violin string that stretches from the coast of South America to the Indonesian archipelago, and when the wind's bow strikes that string, it sets up vibrations—oscillations—that are akin to musical tones.

How old is El Niño? At least two million years old, probably more, according to Jay Noller, a geomorphologist from Vanderbilt University. Noller reached that conclusion after studying the effects of the 1982–83 El Niño on sediments in Peru's coastal desert. In addition to creating a muddy mess, he observed, the rains melded the surface layers of dust into fine red soils that resembled those created by what people in Texas call "mud rain." Along the coast of Peru, in layers exposed by millennia of erosion, Noller found ancient soils of the same telltale composition. After dating these soils, he concluded that El Niño had been operating for a couple of million years and maybe more.

Independently, Michael Dillon, a botanist with the Chicago Field Museum of Natural History, reached a similar conclusion. Without frequent El Niños, he reasoned, the genus of flowering plants called *Nolana* would not have become so commonplace along the desert coast of northern Peru. How long the seeds of *Nolana* can remain viable in desiccated soils is not known, but their period of grace probably does not extend beyond several decades, so that without the pulses of rain brought by El Niño, these highly specialized plants would not likely have worked out the evolutionary strategy they did. Dillon is now pursuing techniques for dating *Nolana*'s DNA that may ultimately reveal just when the El Niño–adapted Peruvian species split off from their ancestral Chilean line. Preliminary evidence suggests that the big event might have taken place as far back as four to eight million years ago, which suggests that the El Niño cycle may be equally ancient.

However, this does not mean that the El Niño cycle has not varied over time, that it has not gone through long periods in which the warm side of the cycle was wimpy and weak, the cold side of the cycle vigorous and strong. The Asian monsoon, for example, weakened quite a bit during the height of the last glaciation some eighteen thousand years ago and strengthened nine thousand years ago, after the world turned balmy. Perhaps the El Niño cycle had done something similar. Perhaps it had entered a super–La Niña state during the Last Glacial Maximum and then tilted in the opposite direction as things warmed up.

THE IDEA THAT the El Niño cycle might vary over time would not have seemed half so interesting had it not been for the fact that, on the cusp of a new millennium, the earth's climate system seemed headed for momentous change. In 1976, it was said, the climate system appeared to cross some invisible threshold—"the magic gate," as University of Arizona paleoclimatologist Julia

Cole termed it—so that from that point on it behaved in detectably different ways. It was around this time that global mean temperatures started to angle steeply up; that the oscillations in the North Atlantic and North Pacific switched to strongly positive modes; and the El Niño cycle seemed to rev into exceptionally high gear.

To Kevin Trenberth, head of the Climate Analysis Section at the Boulder, Colorado–based National Center for Atmospheric Research, these seemingly separate trends gradually started to seem like much more than an extraordinary coincidence. Trenberth was a prominent figure in climatological circles. He had done numerous pioneering studies of the El Niño cycle. He had been among the first to recognize the importance of the decadal-scale oscillation in the North Pacific. He had also called attention to the fact that a powerful La Niña was the driving force behind the 1988 U.S. heat wave and drought that for a time was simplistically blamed on rising greenhouse-gas emissions.

Some may recall that 1988 was the year in which wildfires raged throughout the West, taking down nearly half of Yellowstone National Park. It was also the year in which NASA climatologist James Hansen, on a June day when temperatures soared past 100 degrees Fahrenheit in Washington, D.C., stood up in a U.S. Senate committee hearing room to state that the results from his climate modeling experiments suggested that an intensifying greenhouse effect was affecting the earth's climate system in the here and now.

Reluctant though he was to attribute the 1988 heat wave to rising concentrations of greenhouse gases, Trenberth took the issue of global warming extremely seriously. In late 1995, he and some two hundred other scientists on the United Nations International Panel on Climate Change endorsed the now-famous statement that "the balance of evidence suggests a discernible human influence on climate." Included in that evidence, Trenberth had come to believe, was a striking change in the behavior of El Niño.

In the spring of 1996, Trenberth, along with a colleague, Timothy Hoar, published a controversial, albeit highly technical, paper in *Geophysical Research Letters*. Global warming due to the buildup of greenhouse gases, they suggested, was beginning to affect the El Niño cycle. As evidence, they singled out the El Niño, or string of El Niños, that showed up in 1990 and lingered for about five years. The odds that such an extended ocean warming would occur for purely natural reasons, Trenberth and Hoar determined through statistical tests, were low: One such event might be expected to come along once every fifteen hundred to three thousand years.

The two scientists then went on to puzzle over the imbalance between El Niño and La Niña that marked the nearly twenty years that had elapsed since 1976. In that time frame, they noted, there had been between three and five El Niños (depending on how one parsed the 1990s warmings) and only two La Niñas. Such an El Niño–rich period was highly unusual, they argued; in this case, they calculated the expected repeat interval to be once every two thousand years. As Trenberth and Hoar saw it, there were two possible explanations, one natural, the other not. "Is this pattern . . . a manifestation of global warming and related climate change associated with increases of greenhouse gases in the atmosphere," they asked, "or . . . a natural decadal-time-scale variation?" The latter, they concluded, was "highly unlikely."

Or was it? For their analysis, Trenberth and Hoar depended on the Southern Oscillation Index, an atmospheric measure of El Niño's swings that went back only to 1882, and the brevity of that record, skeptics noted, seemed like a rather puny anchor for such a sweeping indictment. After all, there had been other El Niños that spanned more than two years, and there had been other El Niños of exceptional strength, including the El Niño of 1876–77, which the Southern Oscillation Index just narrowly missed. As a headline for a *Los Angeles Times* article on the subject put it: "Blowing Hot Air or Not? Nothing is Conclusive, but 2 Scientists Say Extended El Niño May Be Sign Of Global Warming."

Had the 1997–98 El Niño not suddenly reared into view, had global mean temperatures not continued their sharp rise, the charge leveled by Trenberth and Hoar might have faded into oblivion. Instead, it took on a life of its own. El Niño had been acting strangely of late, noted the *Washington Post,* the *New York Times,* the *Los Angeles Times, TIME, Newsweek,* and *U.S. News & World Report,* and scientists themselves seemed increasingly unsure about whether its behavior was natural or not. Reflecting on the stir he had caused, Trenberth later said that he had meant to be provocative, that he had intended to stimulate a discussion that would generate an intense round of further research.

As Trenberth saw it, the western basin of the tropical Pacific is a very special place. Because of wind and current patterns, it gathers in sun-warmed water from all over the Pacific, from the coast of California and the coast of Chile, from the coasts of Mexico and Panama, Ecuador and Peru. The western warm pool, in other words, has the capacity to concentrate and therefore to amplify the effects of even a small rise in global mean temperatures, and so it made intuitive sense that one of the consequences of enhanced greenhouse warming might be a speedup in the charge-recharge cycle of the big flywheel that powers El Niño. The problem was, the climate system is so complicated that it often acts in counterintuitive ways.

What Trenberth and Hoar had done was to issue the equivalent of a criminal indictment; it was now up to the scientific community at large to render a verdict of guilty or not. How unusual was it, really, to have so many El Niños all in a row? To have two extremely powerful El Niños spaced so closely together? Was the seeming increase in the magnitude and frequency of events simply a random fluctuation that meant absolutely nothing? Or was it driven, as Trenberth and Hoar had suggested, by the heat-trapping effects of the rather modest buildup in greenhouse gases that had occurred since the end of the nineteenth century?

There were so many questions, and in order to answer them, scientists needed some way of establishing how El Niño had responded to other global warmings and coolings. That meant trying to extend the instrumental record with so-called proxy records, meaning the records of temperature and precipitation preserved by such things as ice caps, lake mud, marine sediments, tree rings, and corals. By 1997 those records were starting to come in, but—like the cores from Kiritimati—they came in frustrating bits and fragments. The proxy records were like a disjointed series of snapshots from a family album that spanned thousands of years, Trenberth lamented, when what scientists needed was a continuous videotape.

IN THE SUMMER OF 1998, Rick Fairbanks was tinkering around his lab at Lamont-Doherty. He and his colleagues, research chemists Jim Rubenstone and Rick Mortlock, were just then preparing to analyze thimble-size vials filled with samples of Kiritimati coral. Most of the cores were still locked away in a basement storeroom beneath the old mansion known as Lamont Hall, but a few were arrayed on top of a metal worktable. Ivory in color, embellished with delicate designs, they looked rather like a Sumerian king's cylinder seals.

Hard though it was, retrieving the coral cores had been the easy part of the enterprise. Now Rick and his colleagues were struggling to make sense of the samples they'd collected. It was going to take them at least a year, probably longer, just to make a first pass through the lot and probably several years to piece together the story their chemistry told.

There was bad news and good news, Rick said. On the one hand, none of the fragments incorporated a climate record long enough to capture El Niño's swings. On the other hand, each fragment contained nuggets of information about the state of the climate system in which El Niño had been embedded. According to a preliminary peek, mean sea-surface temperatures in the area around Kiritimati were about 5 degrees Fahrenheit cooler in 11,000 B.C. than was the case in the twentieth century.

No, Rick laughed, he had not succeeded in his quest for a living centenarian. He'd come close, though. On the way back from Kiritimati, he related, the *Moana Wave* had set anchor off Fanning Island, and he and University of Hawaii graduate student

Kevin Weng had gone out coral prospecting. They'd spotted a really big *Porites* sitting in no more than 50 feet of water and so dived down, scooped it up in a cargo net, and airlifted it to the surface. At that point, a fairly large and extremely aggressive white-tip shark appeared. It made a lunge at Rick, who gave it a hard smack on the nose with one of his electric-blue flippers. Then a storm blew up. Rick and Kevin scrambled back into the boat, marked their coordinates with a hand-held navigational device known as a GPS, and watched the *Porites* sink out of sight.

CHAPTER **TEN**

When I stand among these mighty Leviathan skeletons, skulls, tusks, jaws, ribs and vertebrae, all characterized by partial resemblances to existing breeds of sea monsters, but at the same time bearing on the other hand similar affinities to the annihilated antechronicle Leviathans, their incalculable seniors; I am borne back, by a flood, to that wondrous period ere time itself can be said to have begun; for time began with man.

HERMAN MELVILLE, *Moby Dick*

THE HOT ZONE

In December 1998, more than eight thousand researchers from disciplines that ranged across the earth sciences flooded into San Francisco's Moscone Center to attend the big end-of-year American Geophysical Union meeting. The meetings were always interesting, but this time an undercurrent of excitement coursed through the street-length corridors and cavernous lecture halls. And no wonder. The 1997–98 El Niño, NASA oceanographer Antonio Busalacchi exclaimed at a press conference, was the first to have been observed in real time from start to finish, and it had produced a lode of data that researchers would be mining for years.

In session after session, scientists flashed up images that

showed how a big blob of warmth had appeared at the surface of the tropical Pacific in 1997 and spread from west to east; how a big rise in sea level had surged across the ocean, slamming into the South American coast; how the tropical Pacific had anticipated the switch back to cool conditions by swinging from blue desert to green bloom very early in 1998. They also flashed up images of the tremendous fires that had blazed in the dipterocarp forests of Indonesia.

These images were rendered from data collected by oceangoing buoys and earth-orbiting satellites, but there were plenty of actual photographs, too, including aerial shots of the big bites ocean waves had taken out of the California shoreline and close-ups of Phyllis and Gary LaCombe's landslide-struck house. By now, the house was gone, and a bulldozer had completely reworked the property so that it looked like a vacant lot. The 1997–98 El Niño had disappeared as well, and the robust La Niña that had risen in its place seemed determined to remind onlookers of the power of the cold side of the cycle. In late October 1998, in fact, just as demolition crews were preparing to rip into the LaCombes' house, wind patterns fostered by La Niña summoned forth a hurricane named Mitch.

Mitch was not a run-of-the-mill hurricane. It was the worst storm to hit the Americas in more than two centuries. In 1780, what came to be known as the Great Hurricane killed at least twenty-two thousand people in the Caribbean. Next in the line of infamy used to be the Great Storm that hit Galveston, Texas, in 1900, claiming eight thousand lives. In 1998, Mitch usurped the Great Storm's place. In the space of three days, some ten thousand people died, and three million were rendered homeless by an unholy combination of tidal surges and torrential rains that touched off hundreds of flash floods and landslides. In Honduras, many of the shanty towns clinging to the hills around Tegucigalpa disappeared under tons of mud and debris; in Nicaragua, villages perched on the steep sides of the Casita Volcano were demolished by a river of mud bearing boulders the size of tanks.

The damage tallies for 1997 and 1998 were just now coming

in. Worldwide, it was estimated, El Niño could be held to account for the loss of some $36 billion worth of property and more than twenty thousand lives. And that calculus did not even include the storms that slammed into China in the late spring and early summer of 1998, a transitional time when El Niño was on the wane and La Niña was on the rise. Those storms caused floods along the Yangtze River that destroyed another estimated $36 billion worth of property, caused another forty-five hundred deaths, and temporarily or permanently displaced more than 220 million people.

The postmortems on the forecasts were likewise just now coming in, and as usual, the failures were starting to look more interesting than the successes. In the United States, the precipitation forecasts for the winter of 1997–98 had been a success. In essence, forecasters had drawn a line through the middle of the country, and the big storm systems had tracked to the south of that line almost as if nature were taking its cue from the forecasters, as opposed to the other way around. The temperature forecasts, however, had been off. The Southeast, for example, was projected to be colder than normal; it wasn't. Moreover, higher-than-normal temperatures had enveloped not just the north-central region, which was expected, but also the entire western two-thirds of the country.

Even more startling were the breakdowns in the forecasts for the Indian Ocean region, as first India, then East Africa, then Australia and southern Africa, chimed in a lot wetter than most climatologists expected. By now, all eyes had been drawn to the overwrought state of the Indian Ocean. Typically, the Indian Ocean undergoes a moderate warming during El Niño events, but rarely more than three-quarters of a degree. In 1997 and 1998, however, surface temperatures soared as much as 5 degrees Fahrenheit over seasonal norms, and the sweltering ocean saturated the overlying air with water vapor, the raw material for making rain.

Whether the causes were natural or not, the earth's climate system at the end of the twentieth century was clearly in a riled-up state; 1990, 1995, and 1997 had each, in turn, been the warmest year on record, and 1998 was about to join the list. The 1990s as a whole would shortly become the warmest decade in a century, and there was now reason to suspect that the 1990s might be the warmest decade in many centuries. At the time of the American Geophysical Union meeting, scientists were in the process of using proxy readings encoded by trees, corals, and ice to push the global temperature record much further back in time. The 1990s, they would soon observe in an article published by *Geophysical Research Letters,* stuck out as the warmest decade, with 1998 the warmest year in at least a thousand years.

One could quibble with many aspects of the analysis, particularly the sparseness of proxy data for the earliest centuries. Even so, there seemed to be very little question that the waning years of the twentieth century had not just been warm, but historically warm, and that the 1997–98 El Niño had played out against that backdrop like an actor across a stage. Some scientists were even starting to suggest that the 1997–98 El Niño might have been a less powerful event had the world been a cooler place. Every El Niño, as climatologist Kevin Trenberth liked to observe, represents a kind of global warming in miniature; the tremendous burst of heat released by the 1997–98 El Niño, for example, caused global mean temperatures to shoot up by almost half a degree Fahrenheit. If rising temperatures due to the accumulation of greenhouse gases had the effect of making El Niños more common, then the potential for unwholesome synergy was clearly enormous.

Between 1945 and 1955, for example, sea-surface temperatures in the eastern basin of the tropical Pacific not infrequently dipped below 72.5 degrees Fahrenheit. In the 1980s and 1990s, however, sea-surface temperatures rarely strayed below 76 degrees. In 1997, then, it was easy to imagine that the ocean and atmosphere did not have to work quite as hard as they had in the middle part of the century to ratchet sea-surface temperatures up.

Scientists were now struggling to put the El Niño cycle in a much wider context than they ever had before. What would happen, they wondered, if large portions of the tropical Pacific warmed even more or, for that matter, cooled down so that more or less warm humid air was pumped into the atmosphere overall? The latter possibility could not be ruled out. There were, in fact, a number of scientists who were starting to think that 1998, like 1976, might represent another kind of "magic gate," in this case the beginning of a decadal-scale flip that would return the eastern basin of the tropical Pacific to a cooler background state.

If that happened, it would not mean that concerns about greenhouse-gas emissions were baseless. It would mean that projections of what the future holds would have to take such flips into a more exacting account. The post-1976 rise in sea-surface temperatures in the eastern basin of the tropical Pacific, for example, had been tied not only to the increase in El Niños during that time period but also to the increase in surface temperatures across Europe and North America. If this region of the ocean should undergo broad-scale cooling, there was a possibility that the pronounced rise in U.S. temperatures would taper off and perhaps even start to trend down.

In order to understand El Niño, scientists had been forced to forge a connection between an atmospheric disturbance over the Australasian region and a curious ocean warming off the coast of South America. Now it was beginning to look as if they would need to forge a connection between El Niño and something even larger. Real as it was, El Niño was rapidly becoming a metaphor for what happened when something big and powerful pushed the button that controls the convection engine positioned in the tropical Pacific. There were other climate cycles that were capable of pushing the same button, foremost among them the cycles that for 2 million years had brought the earth in and out of a succession of ice ages.

Water in all its forms—solid, liquid, and gaseous—regulates a number of important climatological parameters, but its vaporous form is arguably the most potent. Among other things, water vapor is the most important of the greenhouse gases; it also warms the atmosphere directly through the heat-releasing process of condensation. As increasing numbers of scientists saw it, water vapor was the blood that flows through the climate system; the tropical Pacific, its beating heart. When that heart slowed down, the world cooled, and when it sped up, the world warmed.

Was the increase in the number and frequency of El Niños in the 1990s a sign that the heart of the climate system was beginning to beat faster? At the Geophysical Union meeting in 1998, that was the big, unanswered question that flitted, phantomlike, in and out of all kinds of sessions.

OHIO STATE GLACIOLOGIST Lonnie Thompson wove his way through the crowd at the Moscone Center, having drawn a bead on Rick Fairbanks, whose height rendered him conveniently easy to spot. With Lonnie was his wife, Ellen Mosley-Thompson, whose West Virginia drawl and diminutive stature camouflaged her intellectual toughness. Long ago, the two of them had divided up the world of ice, so that Lonnie worked in the tropics and Ellen at the poles. It was an arrangement that they had devised very early in their careers, when their daughter was young, as the seasons for mounting expeditions to the tropics and poles were different enough that one parent or the other could always arrange to be home.

Otherwise, though, Lonnie and Ellen did most everything together. They wrote funding proposals together; they analyzed ice cores together; they coauthored papers together, including a paper that had just appeared in *Science* on the ice cores Lonnie had brought back from Sajama, an extinct volcano in southern Bolivia whose conical, snow-clad peak towers 21,500 feet above sea level.

Lonnie had arrived at Sajama's base in 1997, toward the middle of June, just as the power of the newborn El Niño was becoming apparent. The Aymara Indians, who lived in the volcano's shadow, were not happy when they heard about his plans to drill into the ice. They feared that the activity would disturb Pachamama, the great earth mother, and the *achachilas,* the spirits that live in the high mountains. They themselves took great precautions to avoid giving offense to these powerful beings. They feasted and feted them; they built shrines in their honor. Otherwise who knew what horrible things might happen?

Lengthy negotiations ensued. Among other things, Lonnie agreed to make a donation to the village library and set up a local weather station. He also agreed to hire villagers for whatever odd jobs arose. Then he and his team joined the villagers in an ancient purification rite. The ceremony culminated with the sacrifice of a white alpaca not far from the place where Lonnie had set up his base camp. The elevation of the base camp was a bit over 15,700 feet. A short time later, Lonnie established a second camp at 19,000 feet, and from there, he and his colleagues launched their assault on the ice field that lay atop the summit.

As winds filled with dusty grit sandblasted exposed skin and tore at the seams of down-filled parkas, the scientists cored out ice that went back twenty-five thousand years. It was extremely cold on top of Sajama, so cold that Lonnie's toenails blackened with frostbite and fell off. In August, a short time after returning to Columbus, Ohio, Lonnie boarded a flight for Beijing and headed off to run another ice-coring marathon on the top of Dasuopu, a 23,000-foot Himalayan glacier that lies on the flank of Xixiabangma, the world's thirteenth-highest peak.

As Rick Fairbanks once wryly observed, "Lonnie's war stories are better than mine."

Lonnie and Rick had known each other since 1993. They had met in Panama City at a meeting convened to discuss future projects in paleoclimatology. The price tags for one thing and another were beginning to mount up when Lonnie heard Rick grumble rather loudly, "Sounds like a public-works project to me." Rick had a reputation for being pointed in his comments, which did not always endear him to others. It was a trait that tickled Lonnie, though, even when the darts were aimed at him. Once, he recalled, in the course of giving a presentation, he used the term *Mega-Niño* to describe what, in the past, appear to have been exceptionally strong and long-lived atmospheric derangements due to ocean warmings. "Mega-Niño!" he remembers hearing Rick exclaim. "Sounds like a headline in the *National Enquirer*!"

In 1993, Rick was just on the verge of publishing new data from coral cores that he had retrieved from Barbados in 1988, the same year that the issue of global warming was making its transition from a research problem to one that had enormous political, economic, and social dimensions.

Like the reefs off Kiritimati, the Barbados reefs were composed of fossil corals that had marked fluctuations in sea levels by marching up and down. The first results Rick and his colleagues published, in fact, concerned sea levels. In this region of the tropical Atlantic, at least, the corals showed that the ice melt at the end of the last glacial era had, for the most part, been a gradual affair but with two extremely rapid pulses. The first occurred around fourteen thousand years ago, the second around eleven thousand years ago, and both were rather dramatic. During the second pulse, for example, so much water melted that sea levels rose around Barbados between 7 and 24 feet in the space of a single century.

Even more surprising were the results Rick and his colleagues were getting ready to publish. Around eighteen thousand years ago, they had determined, at the end of the last glacial stage, the water around Barbados had been a good 9 degrees Fahrenheit cooler than today. By about 13,500 years ago, sea-surface temperatures had edged a paltry 2 to 3 degrees Fahrenheit higher. Then, between about 12,000 and 11,500 years ago, temperatures quite suddenly shot up, pulling nearly equal to the present. These results flatly contradicted the view held by many climatologists—that the tropics had not varied much in temperature as the earth entered and exited major periods of glaciation—and in 1994 they dropped on the climatological community with a rather loud thud.

A short time later, Lonnie Thompson went public with a very similar set of results. In 1993 he had mounted an expedition to Huascarán, at 22,000 feet the highest mountain in Peru. To reach the 20,000-foot-high col positioned between the mountain's twin ice-glazed peaks, he and his team had hauled themselves and six tons of equipment across a gaping crevasse, using a 50-foot ladder as a bridge. Every night, as they fell into restless sleep, they could hear the ice crack and groan around them; every day, as they drilled, they could hear the roar of avalanches in the distance.

The weather was often horrendous. During a three-day gale, high winds ripped Lonnie's tent from its moorings and sent it skidding toward a precipice with Lonnie inside. Somehow he had the presence of mind to take his ice ax and jam it through the tent's floor, thereby arresting what might have become a 2-mile sky dive without a parachute. Despite all the difficulties, he and his team managed to drill out two ice cores, each well over 500 feet long, and forty-five days later, they brought their booty down with the help of thirty porters and nearly four dozen burros.

Huascarán's ice, as Lonnie and his colleagues reported in 1995, contained an extremely rich climate record. Low pollen counts and high dust concentrations, for example, strongly suggested that conditions in this part of the tropics had been extremely dry when the higher latitudes were locked in ice. But once again it was the temperature readings that riveted attention. Just like the Barbados corals, Huascarán's ice appeared to indicate that, at the peak of the last glaciation, the tropics had been 9 to 10 degrees Fahrenheit colder than today.

It was an important result. If the tropics had responded to global cooling so strongly, observed Columbia University climatologist Wallace Broecker in the journal *Nature,* then they might respond to enhanced greenhouse warming in an equivalently radical fashion. These "findings," he wrote, "should . . . send a strong warning to those who choose to discount the potential of the ongoing greenhouse buildup. The paleoclimate record shouts out to us that, far from being self-stabilizing, the Earth's climate system is an ornery beast which overreacts even to small nudges."

FAIRLY QUICKLY, the majority of scientists conceded that tropical temperatures, both in the atmosphere and in the oceans, probably were cooler during the last glacial maximum than previously assumed. But how cool was cool? From that point on, the debate segued into a highly technical dispute that, in its entanglements, called to mind medieval theologians' arguments over how many angels could fit onto the head of a pin. In this case, the arguments revolved around fine points of geochemistry, such as whether subtle isotopic shifts in the carbonate skeletons of coral polyps and the crystalline lattices of snowflakes were measuring temperature or a mixture of temperature and precipitation.

As everyone knows, water molecules are composed of two atoms of hydrogen and one of oxygen. Even so, these molecules are not identical; they vary in weight depending on how many extra neutrons their atoms happen to carry. This difference is expressed as an isotopic number, which is the sum of the protons and neutrons in the atom's nucleus. An oxygen atom, for example, always has eight protons, so that the difference between oxygen 16 and oxygen 18 boils down to eight neutrons versus ten. It seems like an inconsequential difference, but it is one that scientists have very cleverly exploited. In the case of both corals and ice, it turns out, the ratio between two oxygen isotopes—oxygen 18 and oxygen 16—serves as a proxy for the mercury in a conventional thermometer.

Converting isotopic readings into temperatures, however, is hardly straightforward. Seawater, for example, invariably contains more oxygen 18 than fresh water, reflecting the greater ease with which water molecules containing the lighter isotope evaporate. For the same reason, the water that falls as rain is always biased toward oxygen 16. Add those two facts together and it is easy to see that different patterns of precipitation translate into seawater that is relatively higher or lower in oxygen 18. Today, for example, surface waters in areas of the ocean where rainfall is sparse contain relatively more oxygen 18 and less oxygen 16 than waters in rainier zones. Overall, however, the oceans contain more oxygen 16 than they did during the last glacial maximum, when so much of the precipitation that fell was locked up as ice.

So what were the Barbados corals recording when they showed

a rise in oxygen 18 during the last glacial maximum? Were they registering a change in temperature or a change in precipitation that affected the abundance of oxygen 18 in the water in which they grew? Rick, at least, was able to bolster confidence in the oxygen 18/oxygen 16 temperature readings by including a reading from another kind of thermometer, one based on fluctuations in the ratio of strontium to calcium. It was another funny little quirk corals had: As temperatures fell, they not only grabbed relatively more oxygen 18, but they also became more likely to latch onto strontium as a substitute for calcium.

Huascarán's ice presented a much more difficult conundrum. Over long time spans, it appeared that the ice cores from the tropics had the same oxygen 18/oxygen 16 profile as ice cores from Greenland and Antarctica, that is, the less oxygen 18 they contained, the colder one could deduce conditions had been. The reason was simple: It takes more thermal energy to evaporate heavier molecules of water than lighter ones, so that rain and snow during the last glacial maximum contained even less oxygen 18 than it does today and more oxygen 16.

The problem was, on seasonal time scales, the readings from tropical ice revealed a pattern that was precisely the opposite, so that low levels of oxygen 18 correlated with warmer rather than cooler temperatures. These low levels of oxygen 18 also, it so happened, correlated with periods of maximum precipitation, so that many wondered if Lonnie had mistakenly identified as swings in temperature what were actually swings in precipitation. The reason? Because oxygen 18 is heavier, water molecules that contain it fall out faster, so that as moisture-bearing clouds move in from the ocean, they become progressively depleted of oxygen 18. And the more it rains or snows, of course, the less oxygen 18 there is. So naturally the question arose: What was Lonnie measuring really—temperature or precipitation?

The question had first come up in 1983, when Lonnie published the results from the Quelccaya Ice Cap. It came up again in 1989, when he produced an ice core record from the Dunde Ice Cap in Tibet that, according to his isotopic analysis, showed that the last half century was the warmest period in the last twelve thousand years. In 1995 it came up again. In fact, it kept on coming up. One had the impression that it would for quite a bit longer.

"People have argued that it's precipitation that drives the isotopic ratio," Lonnie reflected, "and that is true when you talk about the seasons, but by the time you get to glacial stage ice, you're looking at time spans averaged out over several hundred years, and seasonality has nothing to do with it."

If Lonnie took his critics seriously, it was because he himself had once harbored similar views. As time went on, however, he amassed more and more evidence that the oxygen 18/oxygen 16 ratios in his ice were, in fact, recording temperature. One key piece of evidence came from the data his Chinese collaborators had been collecting since around 1960 from high-elevation weather stations across the Qinghai-Tibetan Plateau. Especially intriguing to Lonnie, the oxygen isotopes from the ice surrounding those weather stations showed the same relationship with temperature as his ice cores did.

"But the thing that really sells it to me," he continued, "is that we have ice from Huascarán in a place that's extremely wet today but was extremely dry in the glacial, and now we have ice from Sajama, which is very dry today but wet during the glacial stage. In both cases, it's the same story."

The precipitation swings were certainly dramatic. During the glacial era, Huascarán was dry and Sajama wet, so that to Lonnie it seemed almost as if the climate regime had entered a quasi-permanent La Niña–like state. Then, around eleven thousand or so years ago, the precipitation regime flipped. At that point, Huascarán began turning much wetter and Sajama much drier, a pattern that evoked the short-term swings in precipitation caused by the flips of the El Niño cycle. Dry to wet, wet to dry. To Lonnie, at least, it seemed that on long time scales as well as short these areas had invisible strings that connected their climatological fates.

But even more interesting to him was the fact that the temperature swings the two sets of ice cores recorded were not diametrically opposite but almost exactly the same. Sajama's temperature record was a lot more jittery than Huascarán's, but the gist was unmistakable. Around twenty thousand years ago, both records showed, temperatures reached their nadir, then rose to levels equivalent to present a mere five thousand years later. Around thirteen thousand years ago, both locales—one in Bolivia, the other in Peru—again recorded a sudden plunge into cold followed by an exuberant leap into warmth.

In the rapidly warming world of the late twentieth century, the pronounced climatological shift that occurred around eleven thousand years ago seemed of particular interest. Over the next several thousand years, the dry montane forests on Huascarán's eastern slopes would undertake an upward march in response to warmer and wetter conditions, and a huge lake to the south of Sajama would dry up completely, leaving as a relic the largest salt flat in the world, the Salar de Uyuni. And at the end of this time period, scientists were suggesting, the El Niño cycle underwent a transformation that was every bit as remarkable, though exactly what the transformation was or why it occurred was not exactly clear.

A SPECIAL TIME

In 1980, shortly after receiving his master's degree, University of Maine archaeologist Daniel Sandweiss started excavating a site along the northern coast of Peru. He was rummaging through an ancient garbage heap when he unexpectedly stumbled across the discarded shells of tropical oysters. The find struck him as so strange that he named the site Ostra, which is Spanish for oyster. The surprise was, the mollusks were uniformly warm-water species ill adapted to the present-day chilliness of Peruvian waters. Perhaps, Sandweiss conjectured, the ocean had not been so cold way back then.

Sandweiss and two colleagues floated the idea in a paper published in 1986; when others responded skeptically, they collected more evidence. The sediments in which Ostra's mollusks were embedded, they had determined, were between five thousand and seven thousand years old, so Sandweiss and his colleagues went on to examine shells and fish bones found in sites of similar vintage and other time periods as well. The evidence, they asserted in an article published in 1996, all pointed in the same direction: A major climate swing had occurred around five thousand years ago.

Prior to that time, they postulated, the waters off the northern coast of Peru had not cycled between cold and warm, as they do today. Rather, for several thousand years, the offshore environment seemed to have stayed surprisingly, perennially warm. Tropical species of mollusks and fish dominated coastal sites like Ostra that were occupied by people during that time period. Then, quite suddenly, species equipped to handle long intervals of cold water took over. Depending on how one wanted to look at it, the El Niño cycle had either gone into a several-thousand-year pause or had gotten stuck in the "on," that is, warm, position.

Could that really happen? Could permanent warmth replace the cycling warmth brought by El Niño? A debate erupted. Another team of researchers offered a less radical interpretation of the same evidence. The warm-water species found at Ostra and other sites did not come from the ocean at all, they argued, but from the warm, shallow lagoons and embayments that would have readily formed during this era of dynamically changing sea levels.

According to this interpretation of the evidence, the tropical mollusks at Ostra were likely to be a gift of the warm current associated with El Niño, which has long been observed to sweep fish, crustaceans, and mollusks, along with their eggs and larvae, down from equatorial waters. Those organisms fortunate enough to have ended up in sun-warmed lagoons would have survived and matured into edible size, so that El Niño need not have gone into a lengthy hibernation in order for the shells and bones of tropical creatures to end up in the places they did.

There the dispute rested until Donald Rodbell, a geoscientist from Union College in Schenectady, New York, started analyzing sediments he'd retrieved in 1993 from the bottom of Ecuador's Laguna Pallcacocha. The lake is situated in a gorgeous Andean cirque positioned at an elevation of 13,320 feet. It is very near the coast, practically on the continental divide, and during El Niños, rain clouds regularly roll in, unleashing torrents of rain. The rain destabilizes the steep slopes above the lake, causing powdery debris, deposited by ancient glaciers, to slide into the water.

This landslide debris is light gray in color, easy to distinguish from the usual dark organic muck that settles at the lake's bottom, and what immediately struck Rodbell about the 30-foot length of sediment he and a colleague pulled from the lake's bottom was the alternating pattern of light and dark bands that marked the uppermost portion. They were sitting in an inflatable boat, out in the middle of the lake, puzzling over the pattern without understanding its meaning. "We called them 'zebra stripes,'" Rodbell recalled. At the time, Rodbell was finishing a post-doctoral fellowship at Ohio State. When he got back, he stashed the sediment cores from Lake Pallcacocha in a refrigerator next to the freezer that held Lonnie Thompson's ice cores.

It took four more years for Rodbell to figure out what the zebra stripes meant. He gave his first talks on the subject in late 1997, just as sea-surface temperatures off the coast of Peru and Ecuador were approaching their maximum departures from normal. First of all, he noted, the most recent bands of light and dark nicely correlated with the El Niños known to have occurred in modern times. Second, from about fifteen thousand to seven thousand years ago, the light bands were quite rare, occurring in intervals spaced fifteen to seventy-five years apart. From then on, the light bands showed up more often, until between five thousand and three thousand years ago, they assumed a much looser spacing of about two to eight and a half years apart.

El Niños had not stopped entirely, Rodbell argued, but they

had slowed down during the balmy interlude that marked the early part of the Holocene, which is the term scientists use to describe the past ten thousand years. Indeed, between about eighty-five hundred years ago and five thousand years ago, mean global temperatures may well have been a tad higher than they were at the end of the twentieth century, and they ushered in a warm, moist climate regime that seems to have been almost idyllic in many (though not all) parts of the world. In the tropics, grasslands shrank and rain forests expanded. In the higher latitudes, summers grew warmer, and the growing season lengthened. In Canada's Northwest Territories, boreal forests took the place of ground-hugging tundra. Swamps covered the arid lands of northern Australia. There were lakes filled with fish in the Sahara Desert.

In the later part of the Holocene, the earth cooled, and these Edens disappeared. Haunting rock paintings dating from 7,000 to 5,000 B.C. show the big-game animals ancient Saharan peoples once hunted—hippopotamuses, rhinoceroses, and giraffes—while paintings of slightly later vintage depict herds of cattle roaming a landscape that now consists largely of stony outcroppings and drifting sand dunes. The African and Asian monsoons, which had been extremely vigorous in earlier millennia, weakened, so that rainfall became more variable. Was it just a coincidence that at this moment the El Niño cycle appeared to shift from low gear into high?

AT THE AMERICAN Geophysical Union meeting in December of 1998, Lonnie Thompson presented the Sajama results at a session devoted to climate change at high elevations. He was directly followed by one of Rodbell's collaborators, paleoclimatologist David Anderson of the National Oceanic and Atmospheric Administration in Boulder, Colorado. Anderson gave a talk on the Lake Pallcacocha cores, at the end of which he mentioned, almost in passing, an intriguing possibility that he and his colleagues were considering. The slowing of the El Niño cycle at the end of the Pleistocene and beginning of the Holocene, he said, might have had something to do with a change in the distribution of sunlight across the earth's surface.

This change involved one of the so-called Milankovich cycles, which, as far as the earth's climate system is concerned, are the biggest, most important cycles of all. They are responsible for pulling the earth into ice ages and for pulling it out, and they also exercise enormous influence over the times in between. These cycles are named after Serbian scientist Milutin Milankovich, who worked out the underlying mathematics during World War I, while he was a prisoner of war in an Austro-Hungarian jail. One recurs roughly every 100,000 years, another every 41,000 years, and another on a schedule that varies between 19,000 and 23,000 years.

The first and longest cycle involves a very slight change in the eccentricity of the earth's orbit. Because of the gravitational influence of other planets, the earth's path around the sun is not entirely regular. The course the earth follows, in fact, alternates between a slight ellipse and a near circle, and this very subtly alters the pattern of heating from the sun. When its orbit is almost circular, the earth remains virtually equidistant from the sun year-round, and the amount of solar radiation it receives remains fairly constant from season to season. When the orbit is more elliptical, however, the earth experiences slightly larger seasonal temperature swings.

Also important is the change in obliquity, or tilt, of the earth's axis. Whereas the eccentricity of the orbit determines the overall distance of the earth from the sun at different times of year, the tilt dictates how much of the sun's energy reaches the poles. Because of the earth's changing tilt, the contrast in temperature between the equator and the poles was greater in the year 2000 than it was ten thousand years earlier.

To understand why this is so, it is useful to consider the other planets. The poles of Uranus, for example, dip to a nearly horizontal plane. As a consequence, each pole in turn enjoys what, on another planet, would be considered an equatorial clime. A planet like Jupiter, by contrast, juts quite nearly straight up, so that its poles are always very cold. The earth, by contrast, has a slight, but modest tilt. It spins on its axis at an angle that varies over thousands of years from around 22 to almost 25 degrees. As the tilt decreases, so that the earth stands more upright, the poles receive less sunlight; as the tilt increases, the poles warm.

The third cycle is caused by the earth's precession around its axis of rotation. Precession is just a wobbly dance, like that performed by a spinning top. Currently, for example, the axis that runs through the earth's North Pole points toward the North Star at the end of the handle of the Little Dipper. But around ten thousand or so years ago, the axis pointed in a different direction, toward Vega. What the wobble translates into is a change in the time of year when the earth comes closest to the sun. Currently this occurs in January, when the Northern Hemisphere is tilted away from the sun, but around eleven thousand years ago it occurred in June. As a result, more sun reached the Northern Hemisphere during the summer solstice, less during the winter solstice, so that summers were slightly warmer, winters slightly colder. In the Southern Hemisphere, the situation was exactly the reverse.

In the case of the cores from Lake Pallcacocha, it was the precessional cycle that was being nominated as the source of the slowdown in the El Niño cycle, though just how it applied the drag to the system was far from understood. An attractively simple scenario emerged when a group of climate modelers at Lamont-Doherty, including Mark Cane and his former student Amy Clement, carried out an interesting *What if?* experiment. The modelers took the same basic climate model used to predict El Niño and added to it the precessional changes that were known to have occurred over the past twelve thousand years. Then they ran the model in hindcast mode.

Lo and behold, large El Niños were indeed rare between twelve thousand and five thousand years ago, but not because the eastern basin of the tropical Pacific was significantly warmer. Rather, it appeared that the changing pattern of solar heating translated into an increase in the strength of the trade winds midway through the year. On numerous occasions, the winds would relax, and the ocean would start to warm, but then the winds would pick up again, driving the ocean back into La Niña conditions. According to the model, in other words, the background state of the tropical Pacific had most likely tilted not toward warmth, but toward cold. It was a very small tilt, however, so that tropical mollusks might conceivably have survived at Ostra and other sites.

Over the next several years, paleoclimatologists who work with corals would turn up intriguing lines of evidence that also seemed supportive of a general slowdown in the El Niño cycle during this time period. A fossil coral that thrived in the waters off Papua New Guinea 2,650 years ago, for example, indicated that the interval between one El Niño and another was very similar to the late-twentieth-century frequency. But a fossil coral from Australia's Great Barrier Reef suggested that, when it was growing some fifty-eight hundred years ago, El Niño had failed to show up for at least thirty years.

Other windows on El Niño's past opened up. On the one hand, a modern coral from the central Pacific Island of Maiana showed that, in the middle part of the nineteenth century, when the world was a cooler place, El Niño had slowed down, coming and going every ten to fifteen years. On the other hand, fossil corals from New Guinea indicated that the El Niño cycle had been operating at an average frequency of 2.5–7 years at various times during the hundred-thousand-year course of the last glaciation.

Intriguing as it was, the evidence gathered from the corals could neither confirm nor negate the hypothesis that the El Niño cycle was likely to speed up in both frequency and intensity in a warming world. It did show that the El Niño cycle was sensitive to the background state of the climate system, so that its protean changes of mood reflected not only its own intrinsic volatility but also a fair measure of external prodding. As University of Arizona paleoclimatologist Julia Cole put it, "What the past shows is that El Niño has a much wider range of behavior than we thought."

"Let me give you a tour of the place," said Lonnie Thompson, navigating his way through the narrow corridor of papers and boxes that led from his desk to the door of his office at Ohio State's Byrd Polar Research Center. On a coatrack hung a parka with a fur-lined hood and a pair of snow boots. Lonnie put them on whenever he spent long periods of time in the cold-storage facility where he stashed his ice. Contour maps from Peru and China were pinned to the walls; there was even a map of Mars. Like Earth, Mars had ice that encapsulated records of climate; its ice, Lonnie thought, might contain organic material, possibly even DNA, that would confirm a past history of life.

In the early days, Lonnie had melted his ice cores in the field; now he routinely brought them back in their frozen state. Stashed in tightly sealed, well-insulated tubes—the same tubes the military uses for packing howitzer shells—the ice had always managed to make it back to Ohio State without thawing. Not that there hadn't been close calls. In 1986, boxes containing cores from the Dunde Ice Cap in eastern Tibet had sat for hours in some dark corner of the Beijing airport. It took a series of frantic transpacific phone calls before Lonnie convinced airport officials to mount a search.

The cold-storage facility had twin walk-in freezers where the temperature hovered between minus 30 and minus 40, the point at which Fahrenheit and Celsius cross over. It was so cold that despite a swaddling of insulation around the freezers, a pingo—a mound of soil-covered ice—started to heave up from the ground just below. To keep the pingo at bay, Lonnie had equipped the subfloor of the building with a row of space heaters. Pingos are commonplace in Alaska, where they sometimes destroy houses, but to find a pingo growing in Columbus, Ohio, was decidedly unusual.

By this point Lonnie had become a legendary figure, celebrated as much for the torrents of data coming out of his laboratory as for his feats in the field. For years, Columbia University's Wallace Broecker had noted in 1995, Lonnie had "fought not only the cold condition of his field sites but also the lukewarm reception of many of those in our field (including me)." In Broecker's opinion, the results from Huascarán showed "how the tenacity of a lone scientist moving against the grain of conventional wisdom can alter the course of thinking."

Broecker himself had long championed an explanation for global temperature swings that focused on an oscillation in the flow of heat through the world's oceans, particularly the North Atlantic. He was an influential figure and also among the first to call attention to the fact that the climate system was capable of undergoing rapid and, from a human perspective, frightening changes. When Broecker embraced the idea that the tropics might be an important key to this puzzle, the scientific debate about what drove climate change in the past—and hence what might drive it in the future—widened.

Lonnie, not surprisingly, became a passionate partisan in this debate. "If you look at the earth, you'll see right away it's a sphere," he often observed, "and you'll see right away that only a small fraction of the land area lies at the poles, the heat sink. Fifty percent of the land area lies in the tropics, between 30 degrees of latitude north and south. If you were looking for something that would change global climate in a hurry, where would you look?" It wasn't quite so simple as Lonnie made it out to be, of course. Through the oceans as well as the atmosphere, the earth's climate system was so interconnected that neither Lonnie nor anyone else could rule out the possibility that changes at higher latitudes called the tune for the tropics.

At the time, Lonnie and his colleagues were plodding their way through a section of ice retrieved on the 1997 expedition to the Dasuopu Glacier. They had placed the ice in a plastic box to protect it from picking up dust, and they worked on it in a cold room where the temperature was kept at a chilly 27 degrees Fahrenheit. The ice was about a yard in length and semicylindrical in shape, having been sliced in half lengthwise. The other half of the ice core was in a freezer in Lanzhou, where Lonnie's Chinese collaborators were based. Very clearly visible in the ice were the light bands of dust deposited during the dry season, the thicker, lighter bands that had fallen as snow. It was an exquisite object whose polished smoothness called to mind a transparent stalactite.

This stalactite, though, had made its way down from 23,500 feet on the side of a yak. In fact, it had taken forty-four yaks to bring out eighty-eight ice-packed cartons. That was because each carton accommodated six ice-filled tubes, 528 tubes in all, so that placed end to end, the cores that filled the tubes would have measured 476 meters, or nearly 1,500 feet. A year earlier, Lonnie recalled, he had staged a trial run, bringing a few ice-packed cartons back to the States as excess baggage. When the cartons emerged from the plane's cargo hold, the beagles employed by the U.S. Customs Service to sniff out drugs and food appeared utterly flummoxed. "They just sat down and yowled," said Lonnie. "The only thing I could figure was that those boxes had spent two days on a yak and those dogs had probably never smelled a yak before."

Already Lonnie and his colleagues had sliced off thousands of thin slivers from the Dasuopu core. They were now in the process of performing various analyses. They were counting the particles of dust in the slivers; they were weighing the nitrates and sulfates and chloride salts. And most important of all, they were measuring the relative abundance of oxygen 18 versus oxygen 16.

In order to do so, the scientists first had to turn Dasuopu's meltwater into gas. They did this by placing little bits of ice into glass vials that contained a soupçon of carbon dioxide; then they loaded the vials onto a little platform that vibrated and shook, thereby encouraging the oxygen atoms in the water and those in the gas to trade places. A couple of hours later, they sent the gas through a mass spectrometer, a streamlined machine not much larger than a big-screen TV but a lot more expensive.

What happened inside the mass spec resembled a nanoscale drag race between cars and trucks, the cars being the oxygen 16 atoms and the trucks, the oxygen 18 atoms. In essence, the mass spec accelerated these vehicles down a straight stretch, then forced them around a bend. The trucks, being more unwieldy, spun out first, the cars second. At that point, it was simply a matter of collecting the wreckage and weighing it over and over again. To ensure its accuracy, each measurement was always repeated five times.

By this point, Lonnie and his colleagues had worked their way back to eight thousand years ago in Dasuopu's ice, which meant they still had about a thousand years to go. Already, Lonnie was finding interesting patterns in the data, such as the extreme dryness that characterized the monsoon failure that coincided with the big El Niño of 1876–77. Even more striking was the six-year monsoon failure that overlapped with an unusually extended El Niño that started in 1790 and lasted through 1793. Of all the monsoon failures, noted Lonnie, "that was the biggie. That was the biggest event in five hundred fifty years."

Five hundred fifty years was about as far back as the annually resolved record went, Lonnie noted. "There's another thousand years after that where we have some annual resolution, but it comes and goes. After that it's going to be a decadal record, and then it will be century by century."

Each of the proxy records of climate has strengths and weaknesses. Corals, for example, pick up radioactive uranium and thorium, so that even fossil corals can be quite precisely dated. Their annual banding also remains clearly visible over time. On the other hand, it's rare to find a continuous coral record that extends over a century or two. To cover a thousand or more years in a comprehensive way it was necessary to do what Rick Fairbanks did with his Barbados specimens, which was to stitch together a number of different samples.

Ice, by contrast, can produce continuous records of extraordinary length, and in Greenland and Antarctica, where temperatures are low and snowfall is light, the annual layering can remain legible for tens of thousands of years. "The beauty of ice," Lonnie said, "is it's continuous. It's not like you have a little piece here and a little piece there, as so often you do with corals." There is one problem with ice, though. As the weight of the ice on top presses on the ice below, the bottom layers spread out and, as a consequence, thin. Thinning occurs most rapidly in the tropics, where temperatures are higher and snowfall heavier. Very quickly, not just years and decades but centuries and millennia all blur together.

For Lonnie, this was a problem. At Quelccaya he'd lucked out. He could count the annual rings back to the base of the core. He knew he had a climate record that went back fifteen hundred years. But Huascarán's climate record was annually decipherable for only 275 years, and at its base it was so squashed that a segment less than three-quarters of an inch thick contained a climate record that covered hundreds and perhaps even thousands of years. That meant that Lonnie had to resort to other, much more indirect, methods of dating the layers in his cores, including the dubious enterprise that Rick Fairbanks referred to as "wiggle matching."

For example, one of the ways Lonnie established dates was by looking for sulfate lines deposited by known volcanic eruptions, like the eruption of Huaynaputina in southern Peru in A.D. 1600. Another way he did it was by calculating the rate at which the ice in his cores compressed and deformed. But the main way Lonnie established time was by comparing his records with those extracted from ice cores in Greenland, matching the wiggles—the ups and downs in temperature, in particular—in one set of records with those in the other. That was the reason, in fact, that Lonnie very carefully synced the mass specs in his laboratory with the mass specs employed by laboratories working on ice from Greenland.

Then the ice from Sajama came along. Lonnie and his colleagues had noticed the black fleck in the core as it came up from 300 feet down. As soon as they got the ice back to Ohio State, they pulled out that part of the core and melted it. It was at that point that Lonnie found himself gazing into the bulging eyes of a six-thousand-year-old bug. "I couldn't believe it." He laughed. "At first, I thought it was a shrimp!" The Sajama core contained other nuggets of organic material as well: insect wings and bits of bark from polylepis trees whose descendants grow today on the volcano's lower flanks. Suddenly, Lonnie was able to tie the swings of temperature and precipitation in one of his ice cores to an independently established series of dates.

He could do this because the organic material contained carbon 14, a heavier isotope that plants continuously absorb from the atmosphere while they are alive. After these plants—and the animals that eat them die—the carbon 14 they contain radioactively decays at a rate that can be precisely calculated and converted into time. The bits of bugs and bark embedded in Sajama's ice produced a smattering of dates that were nearly unassailable. There was plant material from three hundred years ago, from a thousand years ago and two thousand years ago. There was wood that was twenty-four thousand years old. There was the six-thousand-year-old bug.

And, of course, there were big blank spaces in between; as usual, nature had been more parsimonious than Lonnie would have liked. For the first time, however, he had firm dates that provided him with calibration points that could be used to anchor yet another method of dating, one that relied on air bubbles trapped in the ice. Over the past twenty thousand years, it was well established, the chemical composition of the atmosphere had undergone significant changes, and Sajama's ice contained air bubbles in places where there was organic detritus and in places where there was not. Once Lonnie showed that the carbon-14 dates from the detritus matched the dates that others had assigned to the atmospheric changes, he could use the compositional fluctuations in the air bubbles as believable markers of the passage of time.

Now Lonnie and his colleagues were avidly scanning the ice from Dasuopu for similarly useful snippets of organic material. The expedition to Dasuopu had taken place during the height of the monsoon season, when the skies swirled with dark, menacing clouds and migrating birds, swept off course by the winds, plummeted onto the snow and quickly froze to death. At the time, the scientists had joked about finding stray feathers and perhaps bits of bone in the ice. So far they hadn't found any.

THE DATA UNFURLED across a long scroll of paper that stretched from one end of a big conference table to the other. Each end of the scroll was secured by a wooden rod or length of plastic tubing, and in between was a spectacular sequence of peaks and valleys that evoked a sense of traveling through a mountainous landscape, except in this case one was traveling not through space but through time. The squiggles on the scroll, Lonnie explained, were drawn by a digital artist; they represented a computerized synthesis of all the measurements he and his colleagues had made. There was a scroll for every ice core they had collected, and Lonnie was in the initial stages of comparing one with another.

By this point, Lonnie had collected ice cores from three peaks in the Andes and three positioned around the Qinghai-Tibetan Plateau, the region that is to the Asian monsoon what the equatorial Pacific is to El Niño. He had ice from Dasuopu in the Himalayas. He had ice from Dunde in Tibet. And he had ice from Guliya in China's Kunlun Shan Mountains, which seemed to float, miragelike, above the drifting sand dunes of the Taklimantan Desert. In the local Turkic language, Taklimantan means "go in and you will never come out." At the base of the 1,000-foot core that they extracted from Guliya, Lonnie and his collaborators found ice that they estimated might be as much as five hundred thousand years old.

The stories about climate change Lonnie's cores told were each a little different from the others, but common themes were starting to emerge. All the cores, for example, showed sharp oscillations of temperature and precipitation across a range of time spans, from years to decades to centuries. Now, as Lonnie compared one data set to another, the sawtooth zigs and zags of one mountainous landscape mirrored those of another, or so it seemed to him.

Placed side by side, for example, the climate record from the
Quelccaya Ice Cap in Peru and the Dunde Ice Cap in Tibet revealed
decadal-scale shifts that seemed eerily similar. And there was,
Lonnie thought, a reason. Quelccaya, he noted, borders the Altiplano, the second highest plateau in the world; Dunde, the highest. During El Niños they both tend to turn dry, and during
La Niñas, they both tend to turn wet. Now it looked as if they
might be bound together on even longer time scales, which was
the reason Lonnie so often referred to the patterns in his data as
El Niño– and La Niña–like.

Just as they do on El Niño time scales, Lonnie noted, the precipitation regimes around the world's two highest plateaus appear
to have made parallel swings from wet to dry and dry to wet on
much longer time scales as well. Like Sajama in Bolivia, for example, Dunde appears to have been much wetter during the
glacial era and much drier during the Holocene. "This is an oscillation," Lonnie said. "This is an oscillation on millennial time
scales that's virtually the same as the oscillation on the El Niño
time scale."

What El Niño liked to do, Lonnie was becoming increasingly
convinced, the climate system liked to do, which was to swing
back and forth between extremes on every time scale imaginable.
His tropical ice cores, for example, all registered distinct centuries-long swings in temperature and precipitation that seemed
to him, at least, to coincide with the peculiar climatological intervals known, in Europe, as the Medieval Warm Period and the Little Ice Age. The cores that extended over longer time spans also
registered the sharp swing toward cold that occurred some
13,500 years ago. That swing, Lonnie believed, marked the start
of the short but striking geological interval known as the Younger
Dryas.

Like the Oldest and Older Dryas, the Younger Dryas is named
after a pretty alpine wildflower, white with a yellow center, whose
pollen has long served as a marker of glacial advance. During the
Younger Dryas, temperatures that had been steeply rising underwent a precipitously steep decline. Then, at the end of the

Younger Dryas, around eleven thousand years ago, temperatures warmed again. Both the cooling and the warming that followed were impressively rapid. According to ice cores from Greenland, the cooling that ushered in the Younger Dryas took place in the space of a century, with most of it concentrated in just a few decades. At the end of the Younger Dryas, temperatures in central Greenland shot up by 15 degrees Fahrenheit in the space of a decade or less.

Moreover, in the Northern Hemisphere, at least, the Younger Dryas does not appear to have been unique. The long cores scientists have retrieved from the Greenland ice sheet reveal a succession of similarly strong pulses from cold to warm and warm to cold that recurred at roughly fifteen-hundred-year intervals. And there may even have been a pattern within the pattern. According to one scenario, the first warm pulse in a series tended to be very warm, the next two or three or four successively less so. Then, seemingly out of nowhere, came a tremendous icy blast, accompanied, many scientists think, by surging armadas of ice that ended up in the North Atlantic. After that, a strong warming trend took over, so that the system, in essence, continuously pushed its own reset button.

What could cause such tremendous temperature swings? For a long time, everyone's favorite candidate was what Lamont-Doherty's Wallace Broecker has termed "the great conveyor belt," a system that links the currents in all the world's oceans. In the North Atlantic, the conveyor belt includes water that feeds the Gulf Stream, and the speed at which it runs determines how efficiently warm water is shipped from the equator to the poles. When the conveyor belt slows, northern latitudes grow chilly; when it speeds up, they warm. The slowdowns and speedups are due to variations in the rate at which the water sinks at the far end of its journey, pulling the Gulf Stream behind it. The faster water in the conveyor belt falls to the bottom of the Labrador or Norwegian Seas, the faster the Gulf Stream runs.

At its core, the oscillation in the speed of the conveyor belt is

thought to be due to sharp swings in the saltiness and freshness of
seawater. Salty water is much denser than freshwater, so as it
cools on its journey north, it is more likely to make a direct, deep
dive than spend a lot of time muddling around on the surface.
This is why the great flotillas of ice released from the Hudson Bay
area are thought, by some scientists at least, to have been so important. As they melted, they would have flooded the North Atlantic with big pulses of freshwater that greatly slowed the
conveyor belt. That or something like it—perhaps a flood triggered when an ice dam collapsed—may have caused the Younger
Dryas.

It made for a compelling, self-consistent story, and insofar as
the North Atlantic went, there were reasons to think that the
story might be substantially right. The problem was, what happened at high altitudes did not seem able to explain climate
change on a global scale. As Pennsylvania State University
glaciologist Richard Alley observed, "Everyone will tell you that
it's hard for the North Atlantic to tell the Southern Hemisphere
what to do, but it's easy for the tropics. Everything we know is
saying tropics, tropics, tropics. You can run the world out of the
tropics."

Ideas about how the tropics might influence long-term climate change at higher latitudes were not yet sharp; at scientific
meetings they created little flashes like shooting stars. One idea
for regulating the speed of the conveyor belt, for example, invoked an oscillation in the strength of the northeast trade winds
that swoop across the tropical Atlantic. The stronger the trade
winds blew, the more water vapor the atmosphere picked up
and the more saline the water at the sea's surface became. The
process also worked very nicely in reverse: When the trades
slackened, the fresher the water feeding into the conveyor belt
became.

It was an oscillation between the atmosphere and the ocean
that, in some ways, seemed El Niño–like, though it involved not
temperature but salinity. It was yet another way in which a coupling between the tropical atmosphere and the ocean could

produce far-ranging changes that were both dramatic and sudden. Over the coming years different scientists would float different scenarios about how the tropics factored into the large-scale picture, and they would argue over those scenarios endlessly. Once again, it seemed, scientists were trying to feel their way around something whose immensity they had just barely glimpsed— something that, like El Niño, might reside in equatorial waters and from there reach out to perturb the world.

CHAPTER **ELEVEN**

Not too late is it, even now, the third day, to desist. See!
Moby Dick seeks thee not. It is thou, thou, that madly seek-
est him!

HERMAN MELVILLE, *Moby Dick*

BACK TO QUELCCAYA _____

It is August in the year 2000. Lonnie Thompson is standing on the veranda of a remote field station that belongs to the regional utility company Electroperu. In the foreground wind troubles the waters of the small lake that lies behind a low hydropower dam. In the background the setting sun lays a last fleeting touch on the broad white shoulders of the Quelccaya Ice Cap. Then it happens. The Qori Kalis glacier at the ice cap's edge catches the last rays of light and brims over with glistening gold. In Quechua, says Lonnie, the name Qori Kalis means "gilded goblet," and at times like this, it is easy to imagine that some mountain god is, in fact, lifting Qori Kalis to his lips and drinking in its exotic elixir.

The moment passes. Venus appears in the cobalt sky, heralding the onset of evening. Below the ice cap the world dissolves into gloom. With astonishing rapidity the air sheds its warmth, and with it, the capacity to hold moisture, so that the simple act of breathing produces streamers of mist. Lonnie and his companions retreat indoors. Still wearing their jackets, they huddle around a simple wooden table and sup on soup and bread. Then they turn in, piling on thick layers of blankets. At an elevation of 16,000 feet, nighttime temperatures can be relied upon to drop below freezing.

The next morning, as the sun starts its slow pan across frost-cloaked grass, Lonnie ventures out to the veranda again, and this time he peers at the ice cap through a pair of binoculars. Quelc-caya's surface, he can immediately see, is no longer smooth but ripples with wavy indentations. Equally disturbing are the black fingers of rock intruding into the ice, and though Lonnie can't see them, he knows that lakes are nibbling at the edges.

From looking at satellite images, Lonnie thinks that a lake may have formed near the top of the ice cap as well, and if that has in-deed happened, then the ice cap could dwindle into nothingness in an alarmingly short period of time. That's because, unlike snow and ice, water on balance does not reflect solar energy but ab-sorbs it. A lake in the middle of an ice cap is like a space heater in the middle of an igloo.

After breakfast, Lonnie and the others pile into two dust-covered four-wheel-drive trucks and head for the village of Finaya, a cluster of whitewashed buildings that lies en route. Everyone in the village, it seems, has turned out in greeting, and for a long while a few dozen people just stand there, staring quizzically at their visitors. The men are clad in woolen ponchos and *cholos,* woolen caps knitted in intricate geometric designs. The women are wearing multiple skirts dyed bright primary colors and tall top hats fashioned from soft brown felt. Behind their backs, their glossy black hair falls to their waists in neatly plaited braids.

At Finaya, the two Peruvian members of the expedition, civil engineer César Portocarrero and mountain guide Benjamin Vi-cencio, finalize things with the men who have agreed to supply the packhorses. While they're at it, César and Benjamin take the opportunity to negotiate the purchase of a little fresh meat as well. To the other members of the expedition it sounds like a good idea until the source of the meat, a big old ewe with a noble head and a curly coat of wool, is trotted out for closer inspection.

"Do we have a choice?" Lonnie wonders.

"The sheep's had it," University of Kentucky climatologist Keith Mountain replies. "These folks need the meat. They're going to whack the sheep, anyway; we can just take a little part of the whacked thing."

Lonnie gives his assent, thereby sealing the fate of the sheep, and everyone piles back into the trucks again. "I'm just glad Regina isn't here," Lonnie says, referring to his vegetarian daughter. (As it turns out, no one in the expedition ever eats the pieces of whacked sheep. The meat is eventually given away to the men who supply the packhorses.)

Not far from Finaya, in the middle of a sweeping curve, the two trucks abruptly turn off the road and bump slowly down the side of a steeply inclined meadow. At the bottom of the valley runs the Salcca River, which feeds into the Vilcanota River, which, farther down, becomes the Urubamba, which eventually feeds into the Amazon. The trucks roll into the burbling water and make their way to the other side, where the men from Finaya are waiting. With them are seven horses: four roans, a gray, and two whites. As the animals graze and snort and shuffle about, Lonnie and his team unload their supply-filled duffels from the trucks and watch as the caballeros expertly distribute the load. Then they strike off cross-country.

By now Lonnie has so thoroughly memorized the route that he has no need of a map. He walks slowly and methodically across a rock-strewn grassland whose hills and valleys undulate like standing waves in an ocean. Things have changed since the 1983 Quelccaya expedition, he reflects. For one thing, because the road has been lengthened, the hike into the ice cap no longer takes two days; someone in modestly good shape can make it in about five or so hours, though not without gasping for oxygen in the progressively thinning air.

Shortly after Lonnie rounds a small lake, a black dog charges out, then another; their furious barks announce the proximity of a grazing flock of alpacas and sheep. César Portocarrero picks up a couple of small stones and expertly throws them, not to hurt the dogs but to chase them away. On the ridge high above stands a shepherdess who is keeping watch over the flock as well. She is wearing a crimson skirt, and while her eyes track the strangers, her hands manipulate a wooden spindle that winds wool fibers into yarn.

This is a humanized landscape; it has been for several thousand years, ever since people domesticated the llama and alpaca that thrive in these high mountain pastures. As one walks, one encounters isolated stone houses with animal corrals; one stumbles across little dams and canals built to create artificial marshlands. One of the mosslike plants that grow in these marshlands contains high concentrations of silica; alpaca use it as a kind of emery board to file down their ever-growing teeth.

At last, the camp at the base of the ice cap comes into view. There, between the twin moraines, is the kitchen Lonnie and his colleagues constructed out of boulders so long ago. These moraines come from the Little Ice Age, Lonnie says; they mark the place where Quelccaya's terminal glaciers ended back around 1800. There are older moraines, farther down, that bear mute witness to even more dramatic advances of ice. In the early 1970s, when Lonnie first saw Quelccaya, greenhouse warming had not become the overriding concern it is today. "At the time," he reflects, "a lot of people were worried about the onset of a new ice age. Temperatures had been coming down since about the 1950s. I even remember covers of magazines showing palm trees with ice on them."

It's now dusk; the wind is growing sharper and colder. In the distance, lightning flashes and thunder rumbles; storm clouds swirl around the formidable peak of Nevado Ausangate. As quickly as possible, the members of the expedition unpack their gear and pitch their tents. They stretch a bright blue tarp over the boulder kitchen, and beneath it, Keith Mountain whips together a dinner. It's a stew of sorts, made by combining freeze-dried soup with freeze-dried vegetables and chunks of canned meat. Everyone is tired, almost too tired to eat. Under the blue tarp, candles flicker; above it, the ice cap looms like the negative shadow of night.

THE NEXT DAY, Lonnie leads the way to a boulder he's been monitoring for more than twenty years, clambering over rock that, in 1983, was concealed beneath a layer of snow and ice. A short time later, a lake comes into view. Its water is milky looking, filled with particles of pulverized rock that were once embedded in Quelccaya's ice. Such particles are known as glacial flour, and the fact that they are still suspended in the water is a sign that melting is occurring very rapidly. In 1983, this was a very small lake, Lonnie says; in 1995, it was half this size. And that's just the visible part. Lonnie thinks that the lake may extend beneath the ice another thousand feet or so.

Lonnie slowly makes his way along the edges of the lake, past ice caverns that glow electric blue and a field of sharp crystalline spires that mountaineers refer to as penitente. Finally, he reaches the ridge that leads to the boulder. In 1977, Lonnie notes, the boulder was in contact with the ice; now it sits a good quarter mile away, stranded like a whale on a beach of rock. Lonnie snaps a series of pictures of the boulder, using the same Pentax camera with which he had shot pictures of the boulder nearly a quarter century ago.

After lunch, the group heads off in the opposite direction. Lonnie pauses at the place where he shot the picture that graced the cover of *Science* magazine when some of the first, most astonishing results from Quelccaya were published. That was the picture that showed that the ice cap's sheer sides were layered like a wedding cake, with each year clearly marked by a light band of dust. Alas, that exquisite varved face with which Quelccaya once greeted the world is no more; it exists only in Lonnie's memory and his photographic archives.

Lonnie moves on. Contouring across a broad plateau, he finally arrives at the basaltic outcropping that directly overlooks the Qori Kalis glacier. Again, there is a lake at the edge filled with chunks of ice floating in chalky-looking meltwater. There is also rock exposed all along the edges, and the shape of the glacier has changed, Lonnie says. Once convex, it is now concave; once robustly rotund, it now appears sunken and shriveled. The lake is bigger, too. Since the last time he saw it, Lonnie estimates, the lake has very nearly doubled in size.

Ohio State University survey expert Henry Brecher, a longtime colleague of Lonnie's, climbs up on a really big boulder, positions a special surveying camera on top of a steel marker, and prepares to shoot a sequence of photographs. When he returns to Ohio State, he will meticulously compare these shots with those taken from exactly the same spot on previous expeditions. He will then come up with a quantitative estimate of how fast Qori Kalis is retreating. The number will be shocking. The glacier, Henry will find, is now drawing back from its edges at a rate of more than 500 feet a year.

The next day, Lonnie is ready to make the 2.5-mile climb from the base camp to the ice cap's 18,700-foot summit. He is anxious to see if there really is a lake up near the summit. He is also anxious to update his cores by reaming out the top fifteen or so years of snow. So, after breakfast, he and several others make a preliminary foray to the top, porting up big packs filled with equipment. Benjamin and Keith power ahead; Lonnie lags behind. He's just turned fifty-two and sometimes jokes, half-seriously, that he's getting too old for such a physically taxing line of work.

From time to time, Lonnie stops and bends over as if making a deep, formal bow. It's a technique, he explains, for keeping the blood flowing to your head without overtaxing your heart. Slightly more than halfway up, a swirl of clouds settles on the ice cap, compromising visibility. Lonnie calls a halt to the ascent; he's been caught in storms at the top of Quelccaya before, and more than anything, he worries about lightning. It's wiser, he says, to turn back and wait out the weather.

The next day, Lonnie, Keith, and Benjamin set off for the summit again. Midway up they pick up the cached equipment and haul it to the top. There they pull out the hand-operated auger and start to drill into the ice, but then the bit gets stuck. They free the drill by digging a 10-foot-deep snow pit. Then they cache the equipment and make their way down at the nether edge of twilight. There's no sign of a lake, Lonnie says as he enters the circle of light beneath the blue tarp.

That night the Milky Way spills across the sky like an overturned bucket of paint. One can see the deep black Coal Sack; the smudge of light that belongs to our nearest galactic neighbor, the Large Magellanic Cloud near the base of the Southern Cross. And if one were to get up very early in the morning, one might just be able to catch a glimpse of the seven-star cluster known as the Pleiades, which at this time of year lies very low on the northeast horizon.

For centuries Andean shamans have based their seasonal auguries on the Pleiades. In June, if the stars of the Pleiades shine sharp and bright, they say, the next year's potato harvest will be good; if they grow fuzzy and dim, that drought and frost will cut the yield. Surprisingly, these auguries appear to have a scientifically valid basis. The clarity of the sky, a group of researchers recently reported, greatly diminishes during El Niño years, when the danger of drought and frost is highest. The reason: When the eastern basin of the tropical Pacific warms, more water vapor evaporates into the atmosphere, creating a high-altitude haze of cirrus clouds that are too wispy to be visible but substantial enough to blur the stars.

No one knows how old this augury scheme is, but it is quite possible that it, or something like it, was in place in A.D. 1100, when an exceptionally powerful El Niño swept across this same area, greatly compounding the misery of the long-term drought that had locked in a century before. That drought, as noted in an earlier chapter, was one of those extended rhythmic swings from wet to dry that Lonnie had found in Quelccaya's ice a quarter century ago, and it very quickly became the principal suspect in the downfall of Tiwanaku, the great pre-Incan civilization that sprang up some 230 miles to the southwest of Quelccaya, next to the sapphire blue waters of Lake Titicaca.

RISE AND FALL _____

In early September of 1982, shortly after Lonnie returned to Ohio State from his ninth expedition to Quelccaya, a young archaeologist named Alan Kolata found himself standing on top of a ridge overlooking a small village named Chokara. With him was a Bolivian colleague named Oswaldo Rivera. The two archaeologists had asked the leaders of Chokara to accompany them, for Kolata wanted to point out to them what he thought was a breathtaking sight. All around, for as far as the eye could see, were the outlines of ancient *suka kollus,* which in Aymara means raised fields. The *suka kollus,* Kolata believed, were the secret to Tiwanaku's rise, and he was hoping to convince the leaders of this village to resurrect them.

The timing, Kolata later realized, could not have been worse. The auguries for the approaching planting season had been grim, and day after day the sun rose and set in a sky that was completely devoid of moisture-bearing clouds. A crowd gathered on the ridge. There were angry murmurings, suggestions that white-skinned strangers such as Kolata might be responsible for the deepening drought. For a moment, Oswaldo Rivera seriously worried that the villagers might be on the verge of attacking them with stones.

A couple of weeks later, sea-surface temperatures along the South American coast made the enormous upward lurch that announced the onset of the 1982–83 El Niño. Around Finaya in Peru and Chokara in Bolivia, llamas and alpaca grew thin in sun-scorched grasslands, and thousands fell prey to parasites and disease. Even worse, farmers in the area lost half their potato crop to a combination of drought and frost. During the growing season on the Altiplano, clouds are important for two reasons: because they bring rain and because they trap sun-warmed air near the earth's surface at night. That is why, on this high, windswept plateau, frost is frequently the companion of El Niño droughts.

The situation ought to have improved in 1984, when the eastern basin of the tropical Pacific turned cool, but it didn't. In 1985 and 1986, the drought gave way to such torrential rains that Lake Titicaca rose nearly 10 feet in the space of six months. This time, crops were lost because flooding inundated thousands of acres of farmland along the lake's edge. The water came running into the lake from all over the snow-and-ice-clad highlands, including the Quelccaya Ice Cap. For while the streams that course down one side of the ice cap eventually empty into the Amazon, those that flow down the other side make their way into the Rio Grande, which joins the Rio Azangaro, which flows into Lake Titicaca.

In 1986, just as the first El Niño to be predicted by scientists was beginning, Kolata and Rivera somehow convinced the residents of Lakaya, a village not far from Chokara, to take the bold leap back in time. In 1987, drought and frost again decimated the potato crop in the Altiplano, but not the potatoes in the raised fields the Lakayans had re-created along the shores of Lake Titicaca at a place called Pampa Koani. Under extremely adverse circumstances, these fields produced an astounding 15 metric tons of potatoes per hectare (2.5 acres), compared to the pitiful 1.5 metric tons per hectare the Lakayans harvested from their regular fields.

The word spread. People came from all around to see what seemed like a miracle. That was when it became clear how incredibly ingenious Tiwanaku's agricultural system had been. These ancient people, it seemed, had invented ways to outwit the sharp swings of the El Niño cycle.

The Altiplano, or high plain, which surrounds Lake Titicaca is such a cold, arid place that for a long while it was assumed it could never have supported a large concentration of preindustrial people. "This is not prima facie a region for nurturing or sustaining a large population and certainly not one wherein we should expect to find a capital," wrote explorer George Ephraim Squier in 1877, referring to the desolate ruins of Tiwanaku. Squier could not conceive of Tiwanaku as the political, economic, and cultural nexus of a far-flung empire. He thought of it rather as "a sacred spot or a shrine, the position of which was determined by an accident, an augury, or a dream."

Indeed, Tiwanaku is a strange, dreamlike place. Its central feature is a seven-tiered pyramid 60 feet on each side and more than 50 feet tall, and at its base archaeologists have found mummy bundles filled with headless remains suggestive of human sacrifice. In the middle of a temple known as the Kalassasaya one finds the massive stone portal known as the Gateway to the Sun. At the center of the elaborately carved lintel stands Viracocha, the lord of creation. Holding a staff in each hand, he is flanked on either side by kneeling attendants with the faces of humans, falcons, and condors.

There are many versions of the creation story involving Viracocha, all of them slightly different. The gist, however, is the same. Viracocha, it is said, rose from the sacred waters of Lake Titicaca. He, in turn, invoked the sun and the moon and the stars to rise from its shimmering waters. He also made his way to Tiwanaku, where he either sculpted in stone or modeled in clay the various races that make up humankind. Then Viracocha urged living replicas of these figurines to spring forth from caves and rivers and mountains and valleys and spread out to populate the world.

That Tiwanaku had a profound religious function is indisputable, but it is now universally agreed that it was much more than a shrine. Tiwanaku, as Kolata and others have envisioned it, was a resplendent city-state at the hub of an empire whose influence extended well beyond the area under its direct control. Indeed, the Aymara name for Tiwanaku is Taypi Kala, which means the stone in the center, and between A.D. 500 and 1,000, it appears, the agricultural productivity of this austere landscape increased to the point that as many as 40,000 people may have lived in the city of Tiwanaku, another 125,000 along the shores of the lake and adjoining river valleys.

The *suka kollus* these ancient people built were very large structures, sometimes 30 feet across and 6 or more feet high. They were created by digging soil out from the sides of the fields and piling it on top. The excavations then became canals that supplied water during dry times and drained it off during wet times. Some of the raised fields were painstakingly engineered, with a basement of cobblestones covered by clay, perhaps to keep water from the lake from invading the root zone. Above the clay were carefully sorted layers of gravel and sand, which also enhanced drainage, and then, on the very top, a plentiful covering of loam.

The sophistication of the *suka kollus* went far beyond the simple matter of regulating the water supply. Algae, for example, grew in the canals, providing the farmers of Tiwanaku with a regular supply of green manure. The plants growing in the raised fields then served to strip nitrogen from the water, thus acting as a filtration system that kept the nutrients in the green manure from polluting the lake. The canals were also a force for good on bitterly cold nights, as their water stored up solar heat during the day and then slowly released it during the night. At present, it is worth noting, the owners of orange groves in Florida use sprinkler systems powered by electric pumps to ward off frost; the potato farmers of the Altiplano who have revived the *suka kollus* can rely on their canals to do the same job without additional input of energy.

From about A.D. 500 to A.D. 1000, the agricultural system sustained by the *suka kollus* allowed Tiwanaku to expand in wealth, population, and influence. At one point, this ancient city-state had outposts that extended from the humid jungles on its east to the desert coast on its west. These hinterlands were connected to the heartland by llama caravans that plied regular trade routes. Then, however, big trouble set in, and the central hub from which Tiwanaku projected its influence collapsed. Probably there were multiple factors involved, including political and social strife. But Kolata has mounted a powerful argument that climate change was the primary precipitating event, and one of the linchpins in his argument is the record of a prolonged drought that Lonnie Thompson found in the Quelccaya Ice Cap.

According to that record, the decline in precipitation in the Altiplano began around A.D. 1000, then locked in around the year 1100 and lasted for 350 more years. During this time interval, annual precipitation in the Lake Titicaca drainage basin fell by 15 percent, and in response, the level of the water in the lake dropped, at one point by as much as 36 feet. That this actually happened is borne out by sediment cores extracted from the lake by two teams of paleoecologists who participated in Kolata's project.

As rivers and streams became mere trickles and underground springs dried up, the *suka kollus,* an engineering marvel under different circumstances, could no longer supply enough food for Tiwanaku's now large population. The *suka kollus* could withstand the sharp, punctuated fluctuations in precipitation connected with the El Niño cycle, in other words, but not a chronic, long-lasting shortage of rain.

No one really knows just how Tiwanaku finally fell. Some have invoked warfare as the ultimate instrument of decline. Kolata, however, thinks that as the screws of drought tightened, the residents of the city simply walked away. Sometime around the year 1100, he has imagined, the shamans conducted a ritual that sealed Tiwanaku's demise. Probably they made sacrifices of llamas. Probably they chanted their regrets to Viracocha. Then they left.

Just what happened to the inhabitants of Tiwanaku no one really knows. Some of them no doubt emigrated into the warm, moist *yungas* on the edge of the Amazon, others into the highlands, near the headwaters of snow-fed rivers and streams. But many, Kolata thinks, simply stayed in the vicinity of the great city, eking out a meager existence in the shadow of its ruins. As Kolata, a professor at the University of Chicago, sees it, "The people [of Tiwanaku] did not perish en masse, but their social organization, their economic power, their cities studded with monumental displays of former glory, all these went the way of the dinosaur."

OF ALL THE CHARTS Lonnie Thompson has pieced together, one stands out as endlessly evocative. This chart plots the climate swings recorded by Quelccaya's ice against an archaeological time series of the rise and fall of pre-Incan civilizations. To Lonnie, the degree of overlay has always seemed extraordinary, with coastal cultures rising to preeminence during one of the precipitation regimes captured by the ice and highland cultures during another.

Lonnie thinks that these cultures were picking up the background beat of long-term El Niño– and La Niña–like swings in precipitation. Today, he notes, the Altiplano becomes dry during El Niños and wet during La Niñas, whereas the coastal desert to the north shows the reverse pattern. Assuming that the longer-term climate swings recorded by the Quelccaya Ice Cap followed the same rule, then the rise and fall of ancient cultures might well have been paced by El Niño– and La Niña–like oscillations that favored first one region, then the other.

It's a fascinating but speculative hypothesis. Most archaeologists, even those inclined to accord climate a significant role in human affairs, don't buy it. Drawing on other lines of evidence, they argue that at least some of the long-term climate swings in the Quelccaya record were extremely widespread, so that the fluctuations in rainfall recorded by the Quelccaya Ice Cap would have been felt simultaneously in both the Altiplano and along the northern coast.

Southern Illinois University archaeologist Izumi Shimada, for example, has used the Quelccaya record to develop a compelling scenario for the sixth-century fall of the Mochica Empire in the far north of Peru. The short version of what Shimada thinks happened is this: A multidecadal drought, which the ice cap shows began in A.D. 562, combined with a powerful El Niño to cripple the Mochica power center in the vicinity of the modern city of Trujillo. The prolonged dry spell generated severe food shortages that undermined the political legitimacy of the rulers; the El Niño generated massive floods that destroyed the irrigation system needed to grow food in the desert.

If Shimada is right, then what happened to the Mochica is a textbook example of what University of Florida archaeologist Moseley and colleagues, writing for the journal *Andean Past*, put it, a "convergent catastrophe." As Moseley said:

> The nature of convergent catastrophes can be illustrated by an analogy between human disease and natural disasters. Suffered individually a disease or a disaster is generally survived by a healthy population. Yet when a population is first struck by one malady and then is afflicted by a second or more disorders, the likelihood of recovery becomes tenuous. The potency of multiple natural disasters lies in the compound stress they exert upon populations. Some may collapse while others can respond adaptively.

One of the examples of the latter, which Moseley likes to cite, is the agricultural system developed by the Wari, a southern Andean culture that rose to prominence in the seventh century, shortly after the Moche fell on hard times. The Wari constructed and farmed terraced fields on mountainous slopes, and they irrigated those fields with water from canals that tapped secure high-altitude sources of water. "Bringing steep terrain into agrarian production," Moseley observes, "was obviously critical during the great drought and was one of the few ways to compensate for decreased yields in traditional farming areas."

During the great drought that commenced in the eleventh century A.D., history seems to have repeated itself. Over the course of several hundred years, many people appear to have moved out of coastal valleys and into higher terrain, closer to the spring-fed headwaters of rivers and streams. Once again they built canals and agricultural terraces on slopes that were often breathtakingly steep; they also built big stone corrals, a sign that they were relying more heavily on animals as sources of fiber and food. In time, these settlements and the agricultural system that sustained them were also abandoned, and their remains are as haunting in their own way as the ruins of Tiwanaku and the *suka kollus*.

In archaeological circles, it must be acknowledged, the idea that climate can drive cultural rise and fall remains extremely controversial. Those, like Moseley, who champion it are routinely accused of presenting a deterministic vision that fails to take into account such things as historical contingency and the creativity of the human mind. Yet as one contemplated the extraordinary series of natural disasters that occurred during the 1997–98 El Niño, the idea that cultures might be very strongly affected by climate change did not seem to be all that far off the mark. In fact, at the very same time Moseley was talking about convergent disasters in the past, the International Federation of Red Cross and Red Crescent Societies was talking about compound disasters in the present. Compound disasters, stated the *World Disasters Report* published in 1999, occur when catastrophic natural events hit societies that are "already flawed by inherent economic and political fault lines."

One of the examples cited in the Red Cross–Red Crescent report, in fact, was Peru, where flash floods and landslides not only swept away whole villages but also crippled the transportation infrastructure. The transportation troubles, in turn, severely impacted the mining industry, already reeling from the downturn in prices caused by the financial unraveling of Asia. Peru's economy, in fact, seemed to bleed from a thousand cuts. The fishing industry shuddered as silvery schools of anchoveta fled the warming of ocean waters; the tourism industry tightened its belt as foreign visitors, hearing of all the problems, decided to stay away; the retail industry slumped as consumer spending plummeted.

Soon enough the economic crisis in Peru segued into a full-fledged political crisis, which, two years later, would culminate in the downfall of President Alberto Fujimori. The Fujimori government was, in fact, in the process of free fall in August 2000 when Lonnie Thompson and his colleagues revisited Quelccaya. Indeed, what happened in Peru echoed what had happened in Indonesia just the year before, where President Suharto was also forced to resign after losing his popular mandate. In neither instance was El Niño the only factor, or even the most important factor, and yet in some nebulous way the widespread misery it inflicted in 1997 and 1998 deepened the sense of communal malaise that ultimately brought these governments down.

THE OTHER LEVIATHAN _____

The new day dawns blue and bright. Lonnie and the others get ready to climb Quelccaya again. In the boulder kitchen, under the blue tarp, they gulp down coffee and cereal. They dawdle for a bit, then raise their packs to their shoulders and slog off, up and over the big ridge of rock that lies between the campsite and the place where access to the ice cap is easiest. The first section of ice is steep, though not so steep as to require an ice ax, and the higher one goes, the less extreme the angle of incline becomes. Technically there's nothing hard about climbing Quelccaya; in fact, it's like making one's way up a big dollop of vanilla ice cream. It's the altitude that makes it grueling.

One step after another, Lonnie works his way up, pausing often to rest. An hour passes, then two, then three. Eventually, the summit dome comes into view. Keith and Benjamin have already taken the hand-operated auger out of its cache, and they're starting to drill. Lonnie positions himself behind a low snow wall that Keith and Benjamin constructed the other day to serve as a windbreak. He takes out his saw, his plastic Ziploc bags, his orange-jacketed field book, his waterproof marking pen, and gets ready to go to work.

The cores come up in cylindrical chunks. Those that correspond to the 1997–98 El Niño are immediately distinguishable, so endowed with dust and grime that they are a light charcoal gray. Lower down, the cores are simply cloudy looking, covered over with an opaque layer of ice chips. Lonnie measures and examines each one, scribbling a description into the logbook. He finds the dust layers that mark each annual dry season and cuts the ice into pieces. Then he places each piece into a sequentially labeled bag, which he then crimps closed. It's a form of assembly-line work, tedious and exacting, being done at an altitude of 18,700 feet.

Suddenly the drill jams. It jams at the layer of compacting snow just above the ice layer that holds the record of the 1982–83 El Niño. Eighteen years earlier, Lonnie had retrieved the snow that fell during that El Niño; now he wanted to examine a sample of the ice into which that snow had been transformed. For a short while, he, Keith, and Benjamin discuss the options, which boil down to whether to keep on drilling or not. They decide not to. The main purpose of this expedition is not to bring up more ice but to determine how fast Quelccaya is disappearing. They opt to make radar measurements of the ice cap's thickness instead. Then they disassemble the drill and head back down.

It's sunset by the time they step off the ice and onto rock. It's dark by the time they reach camp. Under the blue tarp, in the boulder kitchen, Lonnie takes his place in the same spot he's occupied since the start. It's a corner spot, near the cookstove and away from the wind, and the pile of rocks Lonnie happens to be sitting on is the equivalent of the chair at the head of the family table. It's Lonnie's spot, and no one else ever tries to sit there.

FOR LONNIE IT has been a long journey, not the journey to Quelccaya in the year 2000 but the much longer journey for which Quelccaya served as the launchpad. It has been a journey that has transformed Lonnie from a dispassionate scientist focused on understanding the natural climate system to a committed partisan in the greenhouse-gas debate. Others might sit on the fence about whether or not human emissions of carbon dioxide and other heat-trapping gases are liable to cause serious problems, but Lonnie does not, and the reason is simple. For the past two decades he has been watching Quelccaya waste away before his very eyes.

"It's really Quelccaya," he says, "that's brought the warming home."

Lonnie and his colleagues started monitoring Quelccaya in 1978, and in the beginning, their objective was simply to map what they expected to be a rather complex pattern of ice retreat and advance. By 1990, however, it had become apparent that the ice cap was shrinking back from its edges at the rate of 10–12 feet a year. Then, between 1990 and 1995, the pace accelerated, to 90 feet a year, and between 1995 and 1998, the pace picked up again, to 150 feet a year. Over that time period, Quelccaya had shrunk greatly in area—from 22 square miles to 17—and like an aging dowager, it had also crumpled in height. "Sometimes," says Lonnie, "you get the feeling you could just sit here and watch it disappear."

Nor is it just Quelccaya that's melting. Thirty years ago, Lonnie observes, Venezuela had six glaciers; now it has two; New Guinea's two ice fields are shrinking so fast, they will be gone in a decade; in the Himalayas, Dasuopu's snow cover has diminished by a third since 1989. And Kilimanjaro, to which Lonnie mounted an expedition earlier in the year, has lost more than 80 percent of its permanent snow and ice since the beginning of the century. One of its smaller ice fields has been reduced to a narrow knifelike sliver that stands some 100 feet tall. Lonnie thinks Kilimanjaro will have no ice left in another ten to twenty years. "The earth is warming," he says, "and it's foolish to pretend that it's not."

What Lonnie thinks is happening to Quelccaya is a kind of replay of what happened at the end of the last glacial maximum, when, according to the dominant hypothesis, rising concentrations of carbon dioxide and other greenhouse gases, like methane, amplified the warming that took place as the earth's orbital gyrations brought it into more favorable alignment (from a human perspective at least) with the sun. Just what the precise sequence of events was no one really knows, but what many scientists believe is that the increase in these gases pushed the big button in the tropical oceans that controls the rate at which water molecules evaporate to fill the atmosphere with a heat-trapping shimmer.

Warm water, for example, holds less carbon dioxide than cold water, so that as soon as the earth began to warm, the oceans would have released more carbon dioxide to the atmosphere. This incremental thickening of the greenhouse-gas layer would, in turn, have enhanced the initial warming. At the same time, however, the warming oceans would have started to pump out more water vapor, with virtually identical results. For this reason, it has never been clear the extent to which the increase in carbon dioxide at the end of the last ice age was the cause of the warm-up or its effect. Probably some of both, as cause and effect are so entangled in the climate system that it's hard to separate the two.

What is clear is that a dramatic rise in atmospheric carbon dioxide has now commenced. The current level of 370 parts per million, Lonnie observes, is the highest in at least 420,000 years, which is the amount of time for which scientists have amassed measurements. Even during other interglacial warm periods, the amount of carbon dioxide trapped in the air bubbles of Antarctica's ice never rose much beyond 300 parts per million. A doubling of that level, which is what a number of scientists expect to occur in the twenty-first century, means that concentrations of carbon dioxide could soon reach 600 parts per million. Yes, there have been times in the earth's 4.5-billion-year history when carbon-dioxide levels were higher, Lonnie acknowledges. "The difference is, there weren't any people around then."

UNDER THE BLUE TARP, one of the candles dies, leaving Lonnie's face in a half-moon of dark. Two thousand years ago, he reflects, there were a quarter of a billion people on earth. Four hundred years ago, there were half a billion; two hundred years ago, there were about a billion; a century ago, there were 1.5 billion. And on or about October 12, 1999, a baby born somewhere in the world brought the count to 6 billion and rising. At last, the rate of increase is slowing, but the growth curve still seems a long way from leveling off. By the year 2050 the earth's population of human beings is likely to approach 10 billion.

It seems like such a staggering figure. Among other things, it virtually guarantees that the torrential rains and searing droughts connected with future El Niños and La Niñas will mean still more loss of life and property, still more social and economic disruption. Even now, it is quite clear, the elemental forces of wind and rain cannot rage without running into all kinds of structures that have been placed in their path: highways, natural-gas pipelines, high-voltage towers, bridges, buildings. They cannot rage without running into tens of millions of people living on steep hillsides prone to landslides, in low-lying coastal zones prone to storm surges, along the banks of flood-prone rivers.

The growth in population is diminishing the collective safety margin in other ways. As people cut and burn down forests, as they pave over floodplains, as they allow animals to overgraze fragile grasslands, they are removing vegetative cover that provides protection against extremes of weather. Among other things, trees and shrubs and grasses help stabilize soil, thereby protecting against landslides. They increase the capacity of watersheds to absorb and retain moisture, thereby mitigating the effects of floods and, perhaps even more important, drought. Trees, shrubs, and grasses are active climatological agents; the moisture they give back to the atmosphere generates convective activity, which in turn generates rain.

There is also the matter of the constraints a more crowded world is liable to place on the ability to respond to major shifts of climate. Time and again, when climate has gone seriously sour, people have grimly pulled up stakes and moved on. In the 1930s, in the middle of the Great Depression, people in the United States dispersed en masse from a drought-riven region. As dust drifted like snow across plow-ravaged prairies and billowed into clouds that erased the sun, thousands of destitute families left Texas, Oklahoma, and Kansas and headed West, drawn by the collective hope of finding employment in the irrigated truck farms and fruit orchards of California's central valley. They streamed over the

Sierras like columns of hungry ants, wrote John Steinbeck in *The Grapes of Wrath,* propelled by a desperate search for work, for food.

The drought that triggered the 1930s Dust Bowl, it is worth noting, was mild compared to the multicentury drought that struck the world around A.D. 1100. Tiwanaku was not the only civilization to succumb to that drought; the Anasazi settlement in New Mexico's Chaco Canyon also vanished during that time period, as if into thin air. If a drought of that magnitude and duration were to affect the Altiplano and the American Southwest today, what would happen to the sprawling cities that have grown up in the meanwhile? What would happen to La Paz and Phoenix? Where would the waves of refugees go? .

And then there is the new problem, which is that human beings are becoming a geophysical force as monstrous, in their own way, as volcanoes. They are spewing out all kinds of heat-trapping gases, not just carbon dioxide but also methane and nitrous oxide. The methane comes from the stomachs of cattle and the anaerobic muck at the bottom of rice paddies. The nitrous oxide comes from the nitrogen compounds applied to agricultural fields as fertilizer. At the same time, they're filling the atmosphere with sulfur dioxide and other aerosols that absorb sunlight and bounce it back out to space, thereby serving to cool things down even while greenhouse gases strive to heat things up.

Through their impacts on landscape, people are not only tampering with the hydrologic cycle that controls evaporation and condensation but also with the climatologically potent variable known as albedo, or whiteness. Albedo is a measure of how efficiently different surfaces reflect sunlight; it acts as a direct control of surface temperatures and as an indirect control of precipitation. Albedo is one reason that the jungly places are warm and the ice-covered poles cold; it is one reason that dark green rain forests are wet and light-colored deserts are dry.

Changes in albedo can have local, regional, and even global effects. Indeed, one of the major uncertainties of the twenty-first century concerns the extent to which water vapor is likely to condense into puffy, low-lying clouds whose albedo is extremely high. The blinding white surface of such clouds, no less than fields of snow and ice, represents a negative feedback of the water-vapor system that may help mitigate the effects of carbon dioxide and other greenhouse gases—and in the end, how much temperatures rise in the twenty-first century will be determined by the balance between all the positive and negative feedbacks in the climate system, both those that are natural and those that are not.

Under the blue tarp, candles are burning down to puddles of wax. A passing cloud drops a patter of hail. Outside looms the great tablet of ice on which so much of import has been written. Inside the boulder kitchen a handful of people shiver in the cold and try to peer into an uncertain future.

All anyone can say is that two extremely large and extremely complex systems—the climate system and the human system—seem headed for confrontation, and more than anything else, it is the uncertainty of how each of these systems is likely to react to the other that makes the buildup in greenhouse gases so troubling. No less than the climate system, political, social, and economic systems are inherently nonlinear, and disturbances, no matter what their cause, catch them at different points in their developmental trajectories. Depending on how close to the edge they happen to be, these systems may absorb a big punch without making much ado, or they may react violently to a rather gentle shove.

In both cases, one of the major imponderables is the effect of what, according to some projections, may be an extremely rapid rate of change. Both the climate system and the human system have components that react to change quickly and others that exhibit great inertia. In a situation where change is slow, the difference in reaction time does not matter so much; in a situation where change is fast, however, the difference may set up instabilities that can drive the system in unexpected directions. "We don't understand the natural climate system," Lonnie says, "and we don't understand what we're doing, and we don't understand what the consequences of what we're doing are."

It is humbling to realize how much about the climate system we still do not understand. It is humbling to realize how mightily climatologists are struggling to identify, and to quantify, all the impacts people are having on the climate; how hard they are struggling to create computer models that are accurate enough to produce forecasts of the future; how, in so many ways, the climate system knows more about us than we know about it.

The climate system does not have to struggle, as climate modelers do, with clumsy approximations and interminable calculations. It effortlessly takes into account everything humans are doing everywhere on earth in the same way it takes into account every photon of light that the sun pours on earth; every droplet of water that evaporates from the ocean; every molecule of carbon dioxide that a tree in the tropical rain forest absorbs; every snowflake, every crystal of ice that reflects sunlight; every dark green leaf that absorbs it. It senses us as a diffuse aura of gases and aerosols, as a changing pattern of moist and dry, as a shifting mosaic of light and dark.

And yet the climate system is fundamentally deterministic. In the end, everything it does can be written out as a series of mathematical equations. In principle, the human system is more flexible; it has the capacity to respond to change, even the prospect of change, in ways that are imaginative and innovative. The dramatic tension in the confrontation that looms in the twenty-first century emanates from that essential difference. One system is a slave to the laws of physics; the other is not only capable of understanding those laws but also of altering its course on the basis of reflection and thought. It is capable of doing so but often does not, and the reasons why have sparked debates even more intense than the current debate over climate.

Under the blue tarp it's getting cold. It's time to turn in for the night. Outside, the great white hump of Quelccaya seems to hint at the presence of a much larger, more mysterious, thing. Herman Melville wrote of the white whale:

Dissect him how I may, then, I but go skin deep; I know him not, and never will. But if I know not even the tail of this whale, how understand his head? much more, how comprehend his face, when face he has none? Thou shalt see my back parts, my tail, he seems to say, but my face shall not be seen. But I cannot completely make out his back parts; and hint what he will about his face, I say again he has no face.

THE PACKHORSES are waiting early the next morning. The utensils and cans and candles in the boulder kitchen are packed up; the blue tarp is pulled down. Soon the long procession out begins, first to the rendezvous with the trucks along the Salcca River, then to the little town of Sicuani, then to the former Incan capital of Cuzco, and finally, to Lima. There the North American members of the expedition board a jet that heads north to Miami, following a route that parallels the soaring snow-capped peaks of the Andean Cordillera.

Clearly visible off the right wing of the plane are the twin spires of Huascarán and the col between them where Lonnie once camped. Off the left wing, hidden by clouds, lies the tropical Pacific. Far below, one knows, the waves are reaching up to touch the winds, and the winds are reaching down to touch the ocean. They are murmuring to each other in conspiratorial whispers, urging the great Leviathan to stir from its ever-restless sleep.

ACKNOWLEDGMENTS

After reading this book, those familiar with climate science might be wondering why I happened to focus on these particular individuals as opposed to so many others. All I can say is that it was an arbitrary decision on my part and a necessary decision, I believe, in order to make this sprawling subject accessible to general readers. For this reason, I have come to think of the scientists in these pages in much the same way as they think of climate records encoded by corals and ice—as proxies for all the others who have contributed so much to the understanding of El Niño and the other cycles that control climate change.

Because of its interdisciplinary nature, climatology is not so much a field of science as a universe. The quest to understand El Niño has thus unfolded as an epic, one that has involved hundreds of individuals from dozens of different fields—oceanography, atmospheric science, glaciology, hydrology, virology, ecology, archaeology, and anthropology, to name just a few. In the course of my research, I have benefited from discussions with many scientists, and I would like to extend my thanks to them all. With apologies to those whose names I have inadvertently omitted, I would like to thank the following:

Hartmut Abberger, Rob Allan, Richard Alley, David Anderson, Jim Arthur, Richard Barber, Timothy Barnett, Gerry Bell, Steve Bourget, David Bowman, Henry Brecher, Wallace Broecker, Antonio Busalacchi, Charles Calisher, Mark Cane, Daniel Cayan, Stanley Changnon, Chris Charles, Francisco Chavez, Amy Clement, Kim Cobb, Julia Cole, Lisa Curran, Henry Diaz, Michael Dillon, Robert Dunbar, Elfatih Eltahir, David Enfield,

Richard Fairbanks, Luis Flores, Michael Glantz, Greg Gurri Glass, Lisa Goddard, Johann Goldammer, Nicholas Graham, William Gray, Rick Grigg, Bart Hagemeyer, Steven Hare, Stefan Hastenrath, Anja Hoffman, Ming Ji, Phil Jones, James Joseph, David Keefer, John Kermond, Ali Khan, George Kiladis, Bruce Koci, Alan Kolata, Arun Kumar, John Kutzbach, Chester Langway, Ants Leetmaa, Robert Livezey, Stephen Lord, Roger Lukas, Nathan Mantua, William McCormick, John McGowan, Michael McPhaden, Ronald McPherson, Gerald Meehl, Michael Moseley, Keith Mountain, Antonio Moura, Paul Neiman, Jay Noller, James O'Brien, Benjamin Orlove, Jonathan Overpeck, Robert Parmenter, William Patzert, Ola Persson, C. J. Peters, George Philander, Roger Pielke Jr., Roger Pielke Sr., David Pierce, Stephen Pyne, Marty Ralph, Eugene Rasmusson, Robert Rauber, Andy Reese, Mark Reid, Richard Reynolds, Donald Rodbell, Chester Ropelewski, Anne Russon, Daniel Sandweiss, Ed Sarachik, Gavin Schmidt, Paul Sereno, Izumi Shimada, Gary Simpson, Willie Smits, Adi Susilo, Emily Teeter, David Thompson, Ellen Mosley-Thompson, Lonnie Thompson, Brian (Owen B.) Toon, Kevin Trenberth, Randall Updike, John M. Wallace, Peter Webster, Lisa Wells, Kevin Weng, Tom Wigley, Raymond Wilson, Klaus Wolter, Christopher Woods, Warren Wooster, Klaus Wyrtki, Terry Yates, Stephen Zebiak.

I have drawn on many sources: newspapers and magazines, scientific journals and numerous books. To cite just a few of the latter: My knowledge of weather and weather history was greatly enhanced by *The Weather Book* by Jack Williams and *The Weather Revolution* by Jack Fishman and Robert Kalish. I learned an immense amount about El Niño's past from the informative volume edited by Henry Diaz and Vera Markgraf, *El Niño and the Southern Oscillation*, as well as from *Currents of Change* by Michael Glantz and *El Niño Southern Oscillation & Climatic Variability* by Rob Allan, Janette Lindsey, and David Parker.

While I was in the midst of writing this book, I read and profited from reading a number of recent books about climate and climate change, including *The Two-Mile Time Machine* by Richard

Alley, *Greenhouse* by Gale E. Christianson, *Floods, Famines and Emperors* by Brian M. Fagan, *Is the Temperature Rising?* by S. George Philander, and *The Change in the Weather* by William K. Stevens. I also had the pleasure of delving into various books about South American archaeology, including *Tiwanaku and Its Hinterland* edited by Alan L. Kolata, *Pampa Grande and the Mochica Culture* by Izumi Shimada, and *The Incas and Their Ancestors* by Michael E. Moseley. Above all else, I am grateful to all the scientists who read and critiqued portions of this manuscript at various stages of its evolution, particularly Mark Cane, Lonnie Thompson, and Kevin Trenberth who were kind enough to tackle the whole thing. I am likewise grateful to my colleagues Maggie Sieger and Frederic Golden who performed the same service from the vantage point of general readers.

In addition, I would like to thank the residents of Upper Canyon Three in Rio Nido, California, for sharing their experiences with me. I would like to thank Jessie Haceesa for sharing her memories of her son Hardy.

I would like to thank Curtis Black and Kate Noble for their help in digging up obscure articles from the late-nineteenth and early-twentieth centuries. I would like to thank Judy Jackson and Agnes Summers for their help in proofreading.

I would like to thank Michael Glantz of the National Center for Atmospheric Research for inviting me to the symposium on El Niño in 1997, where the idea of doing this book was born.

I would like to thank my agents Cherie Burns and Todd Shuster for prodding me to write this book.

I would like to thank my first editor at Warner Books, Claire Zion, for her intense interest in the topic and for peppering me with the essential question "Why are you writing this book?" I would like to thank my second editor at Warner Books, Les Pockell, for his constructive criticism and his enduring enthusiasm.

I would like to thank production editor Penina Sacks for her skill in guiding the book through the critical last phases of the preproduction process.

I would like to thank the editors of *TIME* for allowing me the freedom to take on this project and to see it through. I would like to thank my colleagues at *TIME,* particularly those in Asia and South America, for being so cordial and helpful while I was traveling in their territory.

And last but not least, I would like to thank my husband, Thomas Nash, for his many insights and helpful suggestions and, above all, for his unwavering support without which this book would never have been finished.

SELECTED BIBLIOGRAPHY

Allan, Rob, Janette Lindsay, and David Parker. *El Niño Southern Oscillation & Climatic Variability.* Collingwood, Australia: CSIRO, 1996.

Alley, Richard. *The Two-Mile Time Machine: Ice Cores, Abrupt Climate Change, and Our Future.* Princeton: Princeton University Press, 2000.

Arnold, David. *Famine: Social Crisis and Historical Change.* Oxford/New York: Basil Blackwell, 1988.

Bailey, Eric. *The Christmas Island Story.* London: Stacey International, 1977.

Bowen, Mark, "Thompson's Ice Corps." *Natural History* 107, no. 1 (February 01, 1998): 28.

Broecker, Wallace S. "Cooling the Tropics." *Nature* 376 (1995): 212–213.

———. "Will Our Ride into the Greenhouse Future Be a Smooth One?" *GSA Today* 7, no. 5 (May 1997): 1–7.

Bjerknes, Jacob. "El Niño Study Based on Analyses of Ocean Surface Temperatures 1935–1957." *Bulletin of the Inter-American Tropical Tuna Commission* 5 (1961): 217–303.

———. "Atmospheric Teleconnections from the Equatorial Pacific." *Monthly Weather Review* 67 (1969): 163–172.

Canby, Thomas. "The Year the Weather Went Wild." *National Geographic* 152 (December 1977): 798–829.

Cane, M. A., S. E. Zebiak, and S. Dolan. "Experimental Forecasts of El Niño." *Nature* 322, no. 6073 (1986): 827–832.

Cane, M. A., and M. Evans. "Do the Tropics Rule?" *Science* 290, no. 1107 (November 10, 2000): 1107–1108.

Cappella, Chris. "California Officials Prepare for a Season of Soaking Rains." *USA TODAY* (August 19, 1997): 04.D.

Changnon, S., R. A. Pielke, Jr., D. Changnon, R. T. Sylves, and R. Pulwarty. "Human Factors Explain the Increased Losses from Weather and Climate Extremes." *Bulletin of the American Meteorological Society* 83, no. 3 (2000) 437–442.

Charles, C. D., D. E. Hunter, and R. G. Fairbanks. "Interaction Between the ENSO and the Asian Monsoon in a Coral Record of Tropical Climate." *Science* 277 (August 15, 1997): 925–928.

Christianson, Gale E. *Greenhouse: The 200-Year Story of Global Warming*. New York: Walker & Company, 1999.

Clement, A. C., R. Seager, and M. A. Cane. "Suppression of El Niño During the Mid-Holocene by Changes in the Earth's Orbit." *Paleoceanography* 15 (2000): 731–737.

Cole, Julia. "A Slow Dance for El Niño." *Science* 291 (February 23, 2001): 1496–1497.

Curran, L. M., I. Caniago, G. D. Paoli, D. Astianti, M. Kusneti, M. Leighton, C. E. Nirarita, and H. Haeruman. "Impact of El Niño and Logging on Canopy Tree Recruitment in Borneo." *Science* 286 (December 19, 1999): 2184–2188.

DeVries, T. J., L. Ortlieb, A. Diaz, L. Wells, C. L. Hillaire-Marcel; L. E. Wells, J. S. Noller; D. H. Sandweiss, J. B. Richardson III,

E. J. Reitz, H. B. Rollins, and K. A. Maasch. "Determining the Early History of El Niño." *Science* 276 (May 9, 1997): 965–967.

Daubney, R., and J. R. Hudson. "Enzootic Hepatitis or Rift Valley Fever." *Journal of Pathology* 34 (1931): 545–579.

Diaz, Henry, and Vera Markgraf, eds. *El Niño: Historical and Paleoclimatic Aspects of the Southern Oscillation*. New York: Cambridge University Press, 1992, 2000.

Digby, William. *The Famine Campaign in Southern India 1876–1878*. London: Longmans, Green & Co., 1878.

Fagan, Brian M. *Floods, Famines and Emperors: El Niño and the Fate of Civilizations*. New York: Basic Books, 1999.

Fairbanks, R. G. "A 17,000-Year Glacio-Eustatic Sea Level Record 0–17,000 Years Before Present: Influence of Glacial Melting Rates on the Younger Dryas Event and Deep-Sea Circulation." *Nature* (1989): 342, 637–641.

Fein, Jay S., and Pamela L. Stephens, eds. *Monsoons*. New York: John Wiley & Sons, 1987.

Fishman, Jack, and Robert Kalish. *The Weather Revolution: Innovations and Imminent Breakthroughs in Accurate Forecasting*. Cambridge, Mass.: Perseus Books Group, 1994.

Glantz, Michael H. *Currents of Change*. Cambridge/New York: Cambridge University Press, 1996, 2001.

Glantz, Michael H., and Dana J. Thompson. *Resource Management and Environmental Uncertainty: Lessons from Coastal Upwelling Fisheries*. New York: John Wiley & Sons, 1981.

Glantz, Michael H., Richard W. Katz, and Neville Nicholls. *Teleconnections Linking Worldwide Climate Anomalies*. Cambridge/New York: Cambridge University Press 1991.

Glynn, P. W., ed. *Global Ecological Consequences of the 1982–83 El Niño-Southern Oscillation.* Amsterdam/New York: Elsevier, 1990.

Goldammer, J. G., and B. Seibert. "Natural Rain Forest Fires in Eastern Borneo During the Pleistocene and Holocene." *Naturwissenschaften* 76 (1989): 518–520.

Goldammer, J. G., B. Seibert, and W. Schindele. *Fire in Dipterocarp Forests. In: Dipterocarp Forest Ecosystems: Towards Sustainable Management*, A. Schulte and D. Schöne, eds. Singapore/New Jersey/London/Hong Kong: World Scientific, 1996.

Golden, F. "Pardon El Chichón's Dust." *TIME* 120 (July 5, 1982): 70.

Graves, Ralph A. "Fearful Famines of the Past." *National Geographic* 32 (July 1917): 69–90.

Guilderson, Thomas P., Richard G. Fairbanks, and James L. Rubenstone. "Tropical Temperature Variations Since 20,000 Years Ago: Modulating Interhemispheric Climate Change." *Science* 263 (February 4, 1994): 663–665.

Karl, Thomas R., and Kevin E. Trenberth. "The Human Impact on Climate." *Scientific American* 281, no. 12 (December 1999): 100–105.

Kiladis, George, and Henry Diaz. "An Analysis of the 1877–78 ENSO Episode and Comparison with 1982–83." *Monthly Weather Review* 114 (June 1986): 1035–1047.

Kolata, Alan L. *The Tiwanaku: Portrait of an Andean Civilization.* Oxford: Blackwell Publishing, 1993.

———. *Valley of the Spirits: A Journey into the Lost Realm of the Aymara.* New York: John Wiley & Sons, 1996.

Kolata, Alan, ed. *Tiwanaku and Its Hinterland: Archaeology and Paleoecology of an Andean Civilization*. Washington D.C.: Smithsonian Institution Press, 1996.

Livezey R. E., M. Masutani, A. Leetmaa, H-L. Rui, M. Ji, and A. Kumar. "Teleconnective Response of the Pacific/North American Region Atmosphere to Large Central Equatorial Pacific SST Anomalies." *Journal of Climate* 10 (1997): 1787–1820.

Mantua, Nathan J., Steven R. Hare, Yuan Zhang, John M. Wallace, and Robert C. Francis. "A Pacific Interdecadal Climate Oscillation with Impacts on Salmon Production." *Bulletin of the American Meteorological Society* 78 (June 1997): 1069–1079.

Moseley, Michael E., and Carol J. Mackey. "Chan Chan, Peru's Ancient City of Kings." *National Geographic* 143 (March 1973): 318–345.

Moseley, Michael E. *The Incas and Their Ancestors*. London/New York: Thames and Hudson Ltd., 1992.

Murphy, Robert Cushman. *The Bird Islands of Peru*. New York/London: G.P. Putnam's Sons, 1925.

———. "The Most Valuable Bird in the World." *National Geographic* 46 (September 1924): 278–302.

———. "Oceanic and Climatic Phenomena along the West Coast of South America During 1925." *The Geographical Review* 17 (January 1926): 26–34.

Nash, J. Madeleine. "El Niño's Fury." *TIME* 151, no. 6 (February 16, 1998): 66–73.

Nials, Fred, Eric E. Deeds, Michael E. Moseley, Sheila G. Pozorski, Thomas G. Pozorski, and Robert Feldman. "El Niño: The Catastrophic Flooding of Coastal Peru." *Bulletin of the Field Museum* 50, no. 7 (1979): 4–14, 50, no. 8 (1979): 4–10.

National Research Council. *Learning to Predict Climate Variations Associated with El Niño and the Southern Oscillation: Accomplishments and Legacies of the TOGA Program.* Washington, D.C.: National Academy Press, 1996.

————. *Landslides: Investigation and Mitigation.* Washington, D.C.: National Academy Press, 1985.

Normand, C. "Monsoon Seasonal Forecasting." *Quarterly Journal of the Royal Meteorological Society* 79 (1953): 463–473.

Osler, Scott. "El Niño? It's the Least of Our Worries." *San Francisco Chronicle* (December 10, 1997): A3.

Peters, C. J., and Mark Olshaker. *Virus Hunter,* New York: Anchor Books, 1997.

Petit, Charles. "Scientists Fly into the Heart of Huge Storm." *San Francisco Chronicle* (February 3, 1998): A1.

Pezet, F. A. "The Counter-Current 'El Niño' on the Coast of Northern Peru." Report to the Sixth International Geographical Congress. *Geographical Journal* 7 (1896): 603–6.

Pierrehumbert, R. T. "Climate Change and the Tropical Pacific: The Sleeping Dragon Wakes." *Proceedings of the National Academy of Sciences* 97 (200): 1355–1358.

Philander, S. George. *Is the Temperature Rising?: The Uncertain Science of Global Warming.* Princeton: Princeton University Press, 1998.

Pyne, Stephen. *Burning Bush: A Fire History of Australia.* New York: Henry Holt and Co., 1991.

Ramage, C. S. "Prognosis for Weather Forecasting." *Bulletin of the American Meteorological Society* 57 (1976): 4–10.

Rasmusson, E. M., and J. M. Wallace. "Meteorological Aspects of the El Niño/Southern Oscillation." *Science* 222 (December 16, 1983): 1195–1202.

Richardson, Lewis Fry. *Weather Prediction by Numerical Process.* London: Cambridge University Press, 1922.

Rodbell, Donald T., Geoffrey O. Seltzer, David M. Anderson, Mark B. Abbott, David B. Enfield, and Jeremy H. Newman. "An ~15,000-Year Record of El Niño-Driven Alluviation in Southwestern Ecuador." *Science* 283 (January 22, 1999): 516–520.

Ropelewski, C. F., and M. S. Halpert. "Global and Regional Scale Precipitation Patterns Associated with the El Niño/Southern Oscillation." *Monthly Weather Review* 114 (1987): 2352–2362.

Rowell, Andy, and Peter Moore. "Global Review of Forest Fires." WWF/IUCN Report. (December 1999).

Sandweiss, Daniel H., James B. Richardson III, Elizabeth J. Reitz, Harold B. Rollins, and Kirk A. Maasch. "Geoarchaeological Evidence from Peru for a 5000 Years B.P. Onset of El Niño." *Science*, 273 (September 13, 1996): 1531–1533.

Sandweiss, Daniel H., Kirk A. Maasch, and David G. Anderson. "Climate and Culture: Transitions in the Mid-Holocene." *Science* 283 (January 22, 1999): 499–500.

Satterlee, Dennis R., Michael E. Moseley, David K. Keefer, and Jorge E. Tapia. "The Miraflores El Niño Disaster: Convergent Catastrophes and Prehistoric Agrarian Change in Southern Peru." *Andean Past* 6 (2001): 95–116.

Schreiber, R. W., and E. A. Schreiber. "Central Pacific Seabirds and the El Niño Southern Oscillation: 1982 to 1983 Perspectives." *Science* 225, no. 4663 (1984): 713–716.

Sears, Maj. Alfred F. "The Coast Desert of Peru." *Bulletin of the American Geographical Society* 27, no. 3 (1895): 256–271.

Sette, Oscar, and John Isaacs, eds. *Symposium on the Changing Pacific Ocean in 1957–1958. California Cooperative Oceanic Fisheries Investigations Reports VII.* La Jolla, Cal.: California Cooperative, 1960.

Shapiro, Melvyn A., and Sigbjørn Grønås, eds. *The Life Cycles of Extratropical Cyclones.* Boston: American Meteorological Society, 1999.

Sheppard, P. A. "Obituary Notice: Sir Gilbert Walker, C.S.I., F.R.S." *Quarterly Journal of the Royal Meteorological Society* 85 (1959): 186.

Shimada, I., C. B. Schaaf, L. G. Thompson, and E. Mosley-Thompson. "Cultural Impacts of Severe Droughts in the Prehistoric Andes: Application of a 1,500-Year Ice Core Precipitation Record." *World Archaeology: Archaeology and Arid Environment* 22, no. 3 (1991): 247–270.

Shimada, Izumi. *Pampa Grande and the Mochica Culture.* Austin: University of Texas Press, 1994.

Simpson, George. "Sir Gilbert Walker, C.S.I., F.R.S." *Weather* XIV (1959): 67–68.

Singh, Khushwant, *A History of the Sikhs*, 2 vols., Princeton: Princeton University Press, 1963.

Stevens, William K. *The Change in the Weather.* New York: Delacorte Press, 1999.

Stommel, Henry. *Gulf Stream: A Physical and Dynamical Description.* Berkeley: University of California Press, 1965.

Stommel, Henry, and Elizabeth Stommel. *Volcano Weather: The Story of 1816, the Year Without a Summer*. Newport, Rhode Island: Seven Seas Press, 1983.

Taylor, G. I. "Gilbert Thomas Walker." *Biographical Memoirs of the Fellows of the Royal Meteorological Society*. 8 (1962): 167–172.

Thompson, David W. J., and John M. Wallace. "The Arctic Oscillation Signature in the Wintertime Geopotential Height and Temperature Fields." *Geophysical Research Letters* 25, no. 9 (May 1998): 1297–1300.

Thompson, L. G., E. Mosley-Thompson, and Benjamin Morales Arnao. "Major El Niño/Southern Oscillation Events Recorded in Stratigraphy of the Tropical Quelccaya Ice Cap." *Science* 226, no. 4670 (1984): 50–52.

Thompson, L. G., E. Mosley-Thompson, J. F. Bolzan, and B. R. Koci. "A 1500 Year Record of Tropical Precipitation Recorded in Ice Cores from the Quelccaya Ice Cap, Peru." *Science* 229, no. 4717 (1985): 971–973.

Thompson, L. G., E. Mosley-Thompson, M. E. Davis, P-N. Lin, K. A. Henderson, J. Cole-Dai, J. F. Bolzan, and K-B Liu. "Late Glacial Stage and Holocene Tropical Ice Core Records from Huascarán, Peru." *Science* 269 (July 7, 1995): 46–50.

Thompson, L. G., T. Yao, M. E. Davis, K. A. Henderson, E. Mosley-Thompson, P-N. Lin, J. Beer, H. A. Synal, J. Cole-Dai, and J. F. Bolzan. "Tropical Climate Instability: The Last Glacial Cycle from a Qinghai-Tibetan Ice Core." *Science* 276 (June 20, 1997): 1821–1825.

Thompson, L. G., M. E. Davis, E. Mosley-Thompson, T. A. Sowers, K. A. Henderson, V. S. Zagorodnov, P-N. Lin, V. N. Mikhalenko, R. K. Campen, J. F. Bolzan, J. Cole-Dai, and B. Francou.

"A 25,000-Year Tropical Climate History from Bolivian Ice Cores." *Science* 282 (December 4, 1998): 1858–1864.

Trenberth, Kevin E. "The Definition of El Niño." *Bulletin of the American Meteorological Society* 78, no. 12 (December 1997): 2771–2777.

Trenberth, Kevin E., and Timothy J. Hoar. "The 1990–1995 El Niño-Southern Oscillation Event: Longest on Record." *Geophysical Research Letters* 23, no. 1 (January 1, 1996): 57–60.

Tudhope, Alexander W., Colin P. Chilcott, Malcolm T. McCulloch, Edward R. Cook, John Chappell, Robert M. Ellam, David W. Lea, Janice M. Lough, and Graham B. Shimmield. "Variability in the El Niño-Southern Oscillation Through a Glacial-Interglacial Cycle." *Science* 291 (January 25, 2001): 1511–1517.

Wallace, Alfred Russel. *The Malay Archipelago: The Land of the Orang-Otan and the Bird of Paradise.* London: MacMillan and Company, 1869.

Walker, G. T. "Boomerangs." *Nature* 64 (1901): 338–340.

———. "Correlation in Seasonal Variations of Weather, VIII. A Preliminary Study of World Weather." *Memoirs of the India Meteorological Department* 24, no. 4 (1923): 75–131.

———. "Correlation in Seasonal Variations of Weather, IX. A Further Study of World Weather." *Memoirs of the India Meteorological Department*, 24, no. 9 (1924): 275–332.

Webster, P. J., and S. Yang. "Monsoon and ENSO: Selectively Interactive Systems." *Quarterly Journal of the Royal Meteorological Society* 118 (1992): 877–926.

Webster, P. J., A. Moore, J. M. Loschnigg, and M. Lebann. "Coupled Ocean-Atmosphere Dynamics in the Indian Ocean During 1997–98." *Nature* 40 (September 23, 1999): 356–360.

Williams, Jack. *The Weather Book*. New York: Vintage Books, 1997.

Winterhalder, Bruce P., and Thomas R. Brooke. *Geoecology of Southern Highland Peru: A Human Adaptation Perspective*. Boulder, Colorado: University of Colorado, Institute of Arctic and Alpine Research, 1978.

World Disasters Report. Geneva: International Federation of Red Cross and Red Crescent Societies, 1999.

Wyrtki, K. "Fluctuations of the Dynamic Topography in the Pacific Ocean." *Journal of Physical Oceanography* 5, no. 3 (1975): 450–459.

———. "El Niño—the Dynamic Response of the Equatorial Pacific Ocean to Atmospheric Forcing." *Journal of Physical Oceanography* 5, no. 4 (1975): 572–584.

Wyrtki, K., E. Stroup, W. Patzert, R. Williams, and W. Quinn. "Predicting and Observing El Niño." *Science* 191, no. 4225 (1976): 343–346.

Wyrtki, K. "Water Displacements in the Pacific and the Genesis of El Niño Cycles." *Journal of Geophysical Research* 90, no. C4 (1985): 7129–7132.

COPYRIGHT ACKNOWLEDGMENTS

The author is grateful to the following publishers, individuals, and companies for permission to reprint excerpts from selected material as noted below.

The American Geographical Society for permission to reprint excerpts from "Oceanic and Climatic Phenomena Along the West Coast of South America During 1925," by Robert Cushman Murphy, *The Geographical Review* (January 1926) 26–34.

American Geophysical Union for permission to quote from "The 1990–1995 El *Niño*-Southern Oscillation Event: Longest on Record," by Kevin E. Trenberth and Timothy J. Hoar, *Geophysical Research Letters* 23, no. 1 (Jan. 1, 1996): 57–60.

American Meteorological Society, 45 Beacon Street, Boston, MA 02108, for permission to excerpt from "Atmospheric Teleconnections from the Equatorial Pacific," by Jacob Bjerknes, *Monthly Weather Review* 67 (1969): 163–172; "An Analysis of the 1877–78 ENSO Episode and Comparison with 1982–83," by George Kiladis and Henry Diaz, *Monthly Weather Review* 114 (June 1986): 1035–1047; and "Prognosis for Weather Forecasting," by C. S. Ramage, *Bulletin of the American Meteorological Society* 57 (1976): 4–10.

Blackwell Publishing for permission to quote from *Famine: Social Crisis and Historical Change* by David Arnold. Oxford/NY: Basil Blackwell 1988. Copyright © 1988 by David Arnold.

INDEX

ALSO AVAILABLE FROM WARNER BOOKS

THE PROVING GROUND
by G. Bruce Knecht

The legendary annual race spanning over 600 miles from Australia to Tasmania attracted 115 yachts. Among the skippers was Larry Ellison, one of the world's richest men. But catastrophe struck as apocalyptic waves and hurricane-force winds decimated the fleet, taking the lives of six men. Here is the behind-the-scenes story of one of the worst maritime disasters in recent history. Vivid and unforgettable, this epic, true-life drama portrays the valiant rescue effort waged and the survivors who would be forever changed.

"A page-turner."

—*Washington Post*

SEVEN SUMMITS
by Dick Bass and Frank Wells with Rick Ridgeway

Frank Wells was the head of a major motion picture studio. Dick Bass was an energy and resort entrepreneur. In middle age, both men left behind home, family, and successful careers to share an impossible dream. Facing innumerable dangers and obstacles, they set out to be the first to climb the highest mountain on each of seven continents, from McKinley to Kilimanjaro to Everest. A breathtaking, exhilarating story of the human spirit, this is adventure writing at its finest.

"A hell of an adventure. A book for everyone who's ever dreamed of accomplishing great feats."

—Clint Eastwood

BEYOND THE DEEP
The Deadly Descent into
the World's Most Treacherous Cave
by William Stone and Barbara am Ende with Monte Paulsen

Hidden in a remote region of southeastern Mexico is Huautla—possibly the deepest cave complex in the world. In 1994, an elite team of international explorers set out to go farther into the earth than anyone had before. Led by a driven, larger-than-life caving pioneer, the expedition faced some of the most severe, and most deadly, conditions ever encountered. This is their story—a white-knuckle epic of adventure, obsession, and danger that brilliantly illuminates the human desire to go beyond all limits in every way: to go beyond the deep.

"At long last, an epic to stand beside the adventure classics. BEYOND THE DEEP is *Into Thin Air* without the mountain, *The Perfect Storm* without the sea. Expedition leader Stone is Captain Ahab with a Ph.D., Columbus with a rope and dive tank."
—Jeff Long, author of *The Descent*